Y47
5/86

UCE15

OXFORD MATHEMATICAL MONOGRAPHS

Editors

I. G. MACDONALD   R. PENROSE

Nonlinear waves in one-dimensional dispersive systems

BY

P. L. BHATNAGAR

CLARENDON PRESS . OXFORD
1979

Oxford University Press, Walton Street, Oxford OX2 6DP

OXFORD LONDON GLASGOW
NEW YORK TORONTO MELBOURNE WELLINGTON
NAIROBI DAR ES SALAAM CAPE TOWN
KUALA LUMPUR SINGAPORE HONG KONG TOKYO
DELHI BOMBAY CALCUTTA MADRAS KARACHI

© P. L. Bhatnagar 1979

British Library Cataloguing in Publication Data

Bhatnagar, P L
  Nonlinear waves in one-dimensional dispersive
  systems. — (Oxford mathematical monographs).
  1. Wave motion, Theory of 2. Nonlinear theories
  I. Title II. Series
  531'.1133   QA927   79-40272

ISBN 0-19-853531-7

Photoset and printed in India by Thomson Press (India) Limited

FOREWORD

Wave motions have the characteristic property that after a signal is observed at one point, a closely related signal may later be observed at a different point. Sometimes the main difference between the two signals is in their *amplitude*, perhaps because the wave's energy is being spread out over a larger area (or focused within a smaller area). Apart from any such change in amplitude, however, various changes in the *shape* of the wave-form are also possible, and great interest is attached to the mechanisms producing these. Most mechanisms causing wave-forms to change shape can be analysed to advantage in one-dimensional systems. This is because the difficulties of analysis are greatly reduced in such systems, without the most crucial features of those mechanisms being suppressed.

This monograph uses this simplification to give a most valuable introduction to the principal mechanisms that act to change wave-forms. These mechanisms include dispersion, dissipation, and nonlinearity, either separately or in various combinations. The analysis includes, furthermore, the study of those remarkable classes of wave-forms for which the distorting effects of *different* mechanisms exactly cancel.

Professor Bhatnagar was outstandingly well qualified to write this monograph. In it, he leads the reader progressively, by simple stages, through an extensive mass of sophisticated modern material within this intriguing field. The resulting book is a quite admirable introduction to its subject.

I had written the above words before the deeply regretted and untimely death of Professor Bhatnagar on 5 October 1976, when the world of applied mathematics suddenly lost one of its most respected figures. After the shock of this great loss had subsided, I felt anxious to ensure that Professor Bhatnagar's last book would receive the wide circulation that it richly merits. I am deeply grateful to Dr. Phoolan Prasad for his excellent work as editor. Applied mathematicians owe him a great debt for helping to make this important text generally available.

JAMES LIGHTHILL

PREFACE

Mehta Research Institute, in collaboration with the Indian Mathematical Society, conducted a four-week course on 'Hyperbolic Systems of Partial Differential Equations and Nonlinear waves' from 17 May to 15 June 1976 for the benefit of the research workers desirous of taking up this fascinating, as well as useful, field of creative activity. The author gave a series of lectures on some aspects of the nonlinear waves. He mainly concentrated on the *steady* solutions of the celebrated model equations that go by the name of Burgers equation and Korteweg–de Vries (KdV) equation, and on soliton interactions, and on the meaning of group velocity for the nonlinear dispersive waves and more briefly touched upon the general equation of evolution of which the KdV equation is a particular case. Out of the many equations of evolution, which have attracted the notice of the outstanding workers in the field during last two decades, choice fell on the two model equations mentioned above simply because the Burgers equation is the simplest model of a diffusive wave and the KdV equation is the simplest model of a dispersive wave. The latter equation has further become important on account of the solitary wave solution which it admits.

The present monograph, more or less faithfully presents the contents of the lectures by the author with the exception of the appendix to Chapter 1 and the two appendices to Chapter 2 which have been included to make it self-contained as far as possible.

The subject of nonlinear waves is being pursued very actively at present and consequently the lectures had to be generally open-ended. It is hoped that the present monograph will provide a necessary background on the techniques and the subject matter.

The author wishes to acknowledge his gratitude to many mathematicians whose work has made this lecture course possible. Among them, he is particularly indebted to Professors M. J. Lighthill, G. B. Whitham, P. D. Lax, R. M. Miura, J. M. Greene, C. S. Gardner, M. D. Kruskal, T. Taniuti, and C. C. Wei as their outstanding contributions to the subject formed the backbone of the course.

Allahabad   P. L. B.
September 1976

NOTE

Professor Bhatnagar was carefully editing the manuscript of this book, making minor changes and correcting errors, when he suddenly passed away on 5 October 1976. It was then that I took up the preparation of the final form of the manuscript, fully aware that had he done it himself, he would have achieved a higher level of perfection.

The publication of this book has been possible due to the keen interest shown by Professor M. J. Lighthill, Lucasian Professor of Mathematics, University of Cambridge. The Oxford University Press, London, coming to a quick decision regarding the publication, has speeded up the bringing out of this book.

It was possible for me to complete the final form of the manuscript within a short time due to the spontaneous help given by Dr. V. G. Tikekar, Dr. Renuka Ravindran, and Dr. Swarnalata Prabhu, who, like me, are also students of Professor Bhatnagar.

Indian Institute of Science   PHOOLAN PRASAD
Bangalore
August 1977

CONTENTS

1. Linear waves
    1.1. Introduction   1
    1.2. Linear wave equation: wave-terminology   1
    1.3. General linear equation, dispersion relation   4
    1.4. Dispersive waves: group velocity   6
    1.5. General solution of the linear-wave equation   8
    1.6. Propagation of energy in a dispersive wave   11
    1.7. An important kinematical relation   13
    Bibliography   14
    Appendix I Saddle-point method   15

2. Some nonlinear equations of evolution (steady solution)   21
    2.1. Introduction   21
    2.2. Effect of nonlinearity   22
    2.3. Diffusive waves   28
    2.4. Dispersive waves   31
    2.5. Solitary wave: solitons   37
    2.6. Some other equations of evolution exhibiting solitons   38
    Bibliography   41
    Appendix IIA Equations governing duct flow and shallow
                  water waves on uneven bed   46
    Appendix IIB Reduction theory   51

3. Soliton interaction   61
    3.1. Introduction   61
    3.2. Properties of the Schrödinger equation   61
    3.3 Integrals of equation and relationship between
    KdV equation and the Schrödinger equation   65
    3.4. Time-independence of the eigenvalues of the
    Schrödinger equation, determination of
    scattering parameters   67
    3.5. Inverse scattering problem   70
    3.6. Soliton solution of the KdV equation   71
    3.7. Soliton interaction   83
    3.8. Continuous eigenvalues of the Schrödinger operator   88
    Bibliography   88

4. General equation of evolution   91
    4.1. Introduction   91
    4.2. Definitions   94
    4.3. Solitary-wave solution of the general
    equation of evolution   100

CONTENTS

4.4 Application of the general theory to the
   KdV equation   104
4.5. Eigenspeeds of the general solution
   of the KdV equation   109
Bibliography   111

5. Group velocity: nonlinear waves   113
   5.1. Introduction   113
   5.2. Averaging procedure   114
   5.3. Examples   117
   5.4. The Korteweg–de Vries equation   125
   5.5 Group Velocity: dynamical treatment   133
   Bibliography   138
   Author index   139
   Subject index   141

1

LINEAR WAVES

1.1 Introduction

IN this chapter we will discuss some important properties of *linear* waves which are governed by linear equations and which are usually described as having *small* amplitudes, which, in reality, means *infinitesimally small* amplitudes. The purpose of including this chapter in a monograph on nonlinear waves is threefold: (i) to introduce necessary terminology; (ii) to focus attention on some important properties which are necessary to understand the nonlinear-wave phenomena which are determined by nonlinear systems of hyperbolic equations; and (iii) to prepare a background against which the properties of linear and nonlinear waves may be compared and contrasted.

We note that in this monograph we shall generally consider waves in one-dimension so that only two independent variables x and t will occur in our discussion. We shall designate x as the spatial coordinate and t as the time coordinate; this sort of specification permits us to use the well-known terminology associated with waves, such as wavelength, wave number, period, frequency, amplitude, wave velocity, group velocity, etc.

1.2. Linear wave equation: wave terminology

Let us start with the celebrated wave equation

$$\phi_{tt} = c^2 \phi_{xx}, \tag{1.1}$$

where ϕ is some property associated with the wave and c^2 is a positive constant. This equation determines the spatial and temporal evolution of ϕ in a homogeneous, isotropic, and conservative system. In fact, we shall define a *wave in a general way as a temporal and spatial evolution of an entity.*

We can write the general solution of (1.1):

$$\phi(x, t) = f(x - ct) + g(x + ct), \tag{1.2}$$

where f and g are arbitrary functions. The first term in (1.2), as we know, represents a *progressive* wave moving in the positive direction of the x-axis with a constant speed c, while the second term represents a progressive wave moving in the negative direction of the x-axis with the same speed c.

The argument $x - ct \equiv p_f$ of the f-wave is called its *phase*. Similarly, $x + ct \equiv p_g$ is called the *phase* of the g-wave. Evidently, p_f is constant in space-time if $\dfrac{\mathrm{d}p_f}{\mathrm{d}t} = 0$, i.e. if $\dfrac{\mathrm{d}x}{\mathrm{d}t} = c$. Thus, an observer moving with velocity c with the

f-wave will always notice the same phase of the f-wave and, therefore, the same state of wave motion as indicated by the initial value of f. Similarly, an observer moving with velocity $\dfrac{dx}{dt} = -c$ along the g-wave will always notice the same phase p_g or the same value of g with which he started. The above statement gives physical meanings to the terms *phase* and *wave velocity*, also called the *phase velocity*.

In a periodic progressive wave (say when f is periodic function of p_f and $g \equiv 0$), a point where ϕ is maximum is called a *crest* and a point where ϕ is minimum is called a *trough*.

In the language of the hyperbolic partial differential equations to which class (1.1) belongs, we say that the eqn (1.1) admits two real *characteristics* in the (x, t)-plane:

$$\left. \begin{array}{c} C^+ : \dfrac{dx}{dt} = c \\[3mm] \\ C^- : \dfrac{dx}{dt} = -c \end{array} \right\} \tag{1.3}$$

and

Along the first characteristic C^+, $f =$ constant, while along C^-, $g =$ constant. Thus, $f =$ constant and $g =$ constant are the corresponding *compatibility relations*.

We note that the bidirectional propagation of wave represented by (1.1) is not unexpected. The equation is invariant under the transformation:

$$x \to -x, \quad t \to -t. \tag{1.4}$$

The equation is time-reversible and therefore we can study the *future* as well as the *past* of the wave. In contrast, in Chapter 2, we shall discuss *unidirectional* equations of evolution.

Let us now give some particular values to the functions f and g, say

$$\left. \begin{array}{c} f(x - ct) = a \sin(kx - \omega t), c = \dfrac{\omega}{k} \\[3mm] \\ g(x + ct) = 0 \end{array} \right\} \tag{1.5}$$

and

with constant values of ω and k. Then,

$$\phi = a \sin(kx - \omega t) \tag{1.6}$$

represents a periodic progressive wave of *amplitude* a with wave velocity c given by

$$\omega = kc \quad \text{or} \quad \frac{\omega}{k} = c. \tag{1.7}$$

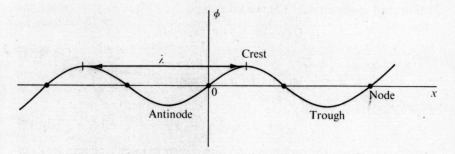

Fɪɢ. 1.1. Plot of ϕ against x for a given value of t.

Eqn (1.6) is a solution of the partial differential equation (1.1) satisfying the initial conditions:

$$\phi(x, 0) = a \sin(kx), \quad \phi_t(x, 0) = -\omega a \cos(kx). \tag{1.8}$$

For a given t, ϕ is sinusoidal in x as indicated in Fig. 1.1.

At any time t, the points $x = (4n + 1)\dfrac{\pi}{2k} + ct$, where $n = 0, \pm 1, \pm 2, \dots$, are where ϕ attains maximum values or crests, and the points $x = (4n + 3)\dfrac{\pi}{2k} + ct$, are where it takes minimum values or troughs. The nomenclature, crest and trough, is derived from the geometric shape of the ϕ-profile in Fig. 1.1. The distance between two consecutive crests (or troughs) is called the *wavelength* and is denoted by λ:

$$\lambda = \left\{ (4n + 5)\frac{\pi}{2k} + ct \right\} - \left\{ (4n + 1)\frac{\pi}{2k} + ct \right\} = \frac{2\pi}{k}. \tag{1.9}$$

From (1.6) it is clear that k gives the number of waves per unit length (taken here in units of 2π) and hence it is called the wave number. All the points on the ϕ-profile at a given time, whose *abscissae* differ by integral multiples of λ, have the same phase.

At a given point with abscissa, say x_1, ϕ oscillates with respect to t with *period*

$$P = \frac{2\pi}{\omega}. \tag{1.10}$$

$\omega = \dfrac{2\pi}{P}$ is called the (angular) *frequency* of the wave and denotes the number of waves passing through a fixed point per unit time (taken here in units of 2π).

If instead of the choice (1.5), we take

$$f(x - ct) = a \sin (kx - \omega t)$$

and   (1.11)

$$g(x + ct) = a \sin (kx + \omega t),$$

we get

$$\phi = [2a \cos \omega t] [\sin kx]$$ (1.12)

so that we can study the variations of ϕ with respect to x and t independently of each other. This choice evidently corresponds to the following initial conditions for ϕ:

$$\phi(x,0) = 2a \sin kx, \quad \phi_t(x,0) = 0.$$ (1.13)

The points $x = \dfrac{n\pi}{k}$, where $\phi = 0$ at *all* times are called *nodes* of the wave. The

points $x = (2n + 1)\dfrac{\pi}{2k}$, where ϕ attains extreme values are called *antinodes*. The

solution (1.12) has been obtained by the superposition of two sinusoidal progressive waves of equal amplitude, wavelength, and frequency, but moving in opposite directions. Except at the nodes, the quantity ϕ oscillates in t with period P, the amplitude at the antinodes is maximum and equal to $2a$ which is clearly equal to the sum of the amplitudes of the component f- and g-waves.

Since there is no communication in the form of energy or momentum transfer between the waves separated by nodes, the wave form represented by eqn (1.12) is called the *standing* wave. The concept of nodes and antinodes are peculiar to the standing wave.

From the above description, it is clear that a solution to (1.1) under certain circumstances represents a *progressive wave* and under certain other circumstances represents a *standing wave*. We also know that the *transverse* wave in an elastic wire stretched taut, in which the motions of various points of the wire are at right-angles to the direction of wave propagation, is represented mathematically by the eqn (1.1). This equation also represents the *longitudinal* sound wave in air in which the air molecules vibrate about their mean positions in the direction of wave propagation.

1.3. General linear equation, dispersion relation

We have introduced the above terms with the help of a specific equation, called the standard linear wave equation. Let us now consider a general linear partial differential equation in two independent variables x and t:

$$L[\phi] = 0,$$ (1.14)

where L is a linear differential operator. When the required initial conditions:

$$\phi(x,0) = \phi_0(x), \; \phi_t(x,0) = \phi_1(x), \ldots, \tag{1.15}$$

are given, we can reduce the given equation to an ordinary differential equation in x-variable with the help of the Laplace Transform technique provided the functions $\phi_i(x)$ are sufficiently smooth. It is always possible (under certain conditions) to solve a linear ordinary differential equation, at least in principle, provided the necessary boundary conditions are given. However, at present we are not interested in this approach as we are interested in a general discussion of eqn (1.14). The equation being linear, we can build up its general solution by superposition of its various Fourier components. Consequently, let us substitute

$$\phi = a \exp\{i(kx - \omega t)\} \tag{1.16}$$

in eqn (1.14) in which we assume now that the independent variables x and t do not appear explicitly and the equation is homogeneous. This substitution removes all derivatives with respect to t and x: $\dfrac{\partial}{\partial t} \to -i\omega, \dfrac{\partial}{\partial x} \to ik$ and reduces it to the following relation:

$$D(\omega, k; A_i) = 0, \tag{1.17}$$

where A_i are the parameters occurring in eqn (1.14). Eqn (1.17) is the *dispersion relation* which determines the frequency ω of the wave in terms of the wave number k and the parameters A_i. We shall write (1.17) *formally* as

$$\omega = \omega(k; A_i). \tag{1.18}$$

The number of roots of eqn (1.17) depends on the degree n of this algebraic equation in ω. Clearly, n is equal to the highest order of t-derivative in eqn (1.14). We consider each root separately as each one of them gives a separate wave, called a *mode*.

Let us consider a general root

$$\omega = \omega(k), \tag{1.19}$$

where we have supressed the dependence of ω on A_i as in the present discussion they do not play any specific role. The corresponding Fourier component is

$$\phi(x,t) \propto \exp[i\{kx - \omega(k)t\}].$$

The temporal evolution of ϕ depends on the nature of $\omega(k)$. The following cases arise:

  (i) when $\omega(k)$ is real, this Fourier component represents a harmonic wave.
  (ii) When $\omega(k)[=i\omega_2(k)]$ is pure imaginary,

$$\phi(x,t) \propto \exp(ikx). \exp\{\omega_2(k)t\}$$

so that we get a nonpropagating standing wave. If $\operatorname{Im} \omega(k) > 0, \phi$ becomes unbounded exponentially with t; if $\operatorname{Im} \omega(k) < 0, \phi$ decays exponentially with t. Thus, in the former case, we deal with a growing (amplifying) wave, while in the latter case we deal with a decaying (attenuating) wave. In the former case, the initial disturbance imposed on the system grows without bound and the system is said to be *unstable* with respect to the particular mode under consideration; in the latter case, the system is said to be *stable* with respect to the mode.

(iii) Let
$$\omega(k) = \omega_1(k) + i\omega_2(k),$$

where ω_1 and ω_2 are real. Here

$$\phi \propto \exp[i\{kx - \omega_1(k)t\}] \cdot \exp(\omega_2(k)t)$$

so that when $\omega_2 = \operatorname{Im} \omega < 0$, the wave is harmonic with exponentially decaying amplitude. The system is stable with respect to the mode in this case. When $\operatorname{Im} \omega > 0$, the wave is harmonic with exponentially growing amplitude. The system is unstable with respect to the mode but Eddington (1926) calls this type of instability *overstability* because it is provoked by restoring forces so strong as to overshoot the corresponding position on the other side of the equilibrium. This sets up an oscillation of increasing amplitude.

The above discussion of the dispersion relation brings out clearly its importance in determining the response of the system to an imposed disturbance, which is assumed to be of infinitesimally small amplitude initially.

The dispersion relation also provides a basis for another classification of waves. Let us *assume* that eqn (1.19) determines a real value of ω for each value of $k: 0 \leq k < \infty$. If $\dfrac{\partial^2 \omega}{\partial k^2} = \omega''(k) \not\equiv 0$, the wave is said to be *dispersive*, when $\omega''(k) \equiv 0$, it is said to be *non-dispersive*. It also introduces a new characteristic velocity, called the *group velocity* denoted by $V_g = \omega'(k)$.

The dispersion relation provides still another basis for classification of waves. When eqn (1.19) determines a complex value for ω, the wave is said to be *diffusive*; when ω is real, the wave is said to be *non-diffusive*. The diffusive waves are associated with attenuation of the amplitudes with time due to certain dissipative mechanisms present in the system.

1.4. Dispersive waves: group velocity

Having introduced the terms group velocity, and dispersive and non-dispersive waves mathematically in an abstract manner, we shall now give them some physical meaning.

Group velocity

Let us consider the superposition of two harmonic waves which differ very slightly in their frequencies and wave numbers, but have the same amplitude:

$$\phi_1(x,t) = a \cos (kx - \omega t), \tag{1.20}$$

$$\phi_2(x, t) = a \cos \{(k + \delta k)x - (\omega + \delta\omega)t\}. \tag{1.21}$$

As a result

$$\phi = \phi_1 + \phi_2 = \left[2a \cos \left\{ \frac{1}{2}(x\delta k - t\delta\omega) \right\} \right] \cos \left\{ \left(k + \frac{\delta k}{2} \right)x - \left(\omega + \frac{\delta\omega}{2} \right)t \right\},$$
$$\tag{1.22}$$

which is the familiar expression for *beats*.

ϕ oscillates with frequency $\omega + \frac{1}{2}\delta\omega$ which is slightly different to ω and has a wavelength which is also slightly different to $\lambda = 2\pi/k$. The effective amplitude

$$A = 2a \cos \{\tfrac{1}{2}(x\delta k - t\delta\omega)\} \tag{1.23}$$

varies slowly with period $\dfrac{4\pi}{\delta\omega}$ and wavelength $\dfrac{4\pi}{\delta k}$ between the sum of the amplitudes of the component waves and zero. Since $\delta\omega$ and δk are small, the period and wavelength of A are both large.

As a result of constructive and destructive interference, the ϕ-profile along both time and space axes appear as a series of periodically repeating groups as shown in Fig. 1.2. Each group consists of a number of waves.

The surface over which the group amplitude remains constant is defined by the equation

$$x\delta k - t\delta\omega = \text{Constant}. \tag{1.24}$$

From eqn (1.24) it follows that the groups themselves are propagated with velocity

$$\frac{\mathrm{d}x}{\mathrm{d}t} = \frac{\delta\omega}{\delta k} = \omega'(k) \text{ (in the limit } \delta k \to 0), \tag{1.25}$$

where the prime denotes differentiation with respect to k. Therefore, the group velocity V_g is given by

$$V_g = \omega'(k)$$
$$= \frac{\text{difference in frequencies of component waves}}{\text{difference in wave numbers of component waves}}.$$

The above discussion imparts physical meaning to the group velocity.

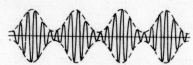

FIG. 1.2. Formation of 'beats' or 'groups' by superposition of two harmonic waves.

We have earlier defined the phase velocity V_p through the relation

$$V_p = \omega/k.$$

Dispersive and non-dispersive waves

In general, both phase velocity and group velocity are functions of the wave number. We can easily check that when $\omega''(k) \neq 0$, V_g is different from V_p and depends on k so that the waves of different wavelengths travel with different group velocities. Let us consider a disturbance initiated at $x = 0$ at time $t = 0$ which consists of a superposition of a number of wavelengths. Since the components of the disturbance with different wave numbers travel with different velocities, after some time the disturbance will be spread over a certain length which increases with time. In this situation we say that the wave has undergone *dispersion*. It is also clear that along the wave-train the wave number varies slowly.

When $\omega''(k) \equiv 0$, both the phase velocity and group velocity coincide and there is no separation of waves of different wave numbers. In this case the wave is non-dispersive.

Example

On substituting (1.16) in (1.1) we get the dispersion relation $\omega = \pm ck$, so that here $V_g = V_p = \pm c$ and the wave represented by (1.1) is non-diffusive and non-dispersive.

1.5. General solution of the linear-wave equation

We have so far considered individual Fourier components of a linear wave. We can obtain the general solution of the equation by superposing these individual Fourier components:

$$\phi(x,t) = \int_{-\infty}^{\infty} A(k)\exp[i\{kx - \omega(k)t\}]dk, \tag{1.26}$$

where $\omega = \omega(k)$ is the function of the wave number and the parameters of the problem as determined by the dispersion relation and the *spectrum function* $A(k)$ takes care of the initial condition. In principle, we can always construct the spectrum function in a given problem though at times it may be very tedious to do so. The solution (1.26) corresponds to the initial condition

$$\phi(x,0) = \int_{-\infty}^{\infty} A(k)\exp(ikx)dk, \tag{1.27}$$

which is the Fourier integral for $\phi(x,0)$ and consequently, given $\phi(x,0)$, $A(k)$ can be evaluated.

We shall now discuss the asymptotic behaviour of eqn (1.26) as $t \to \infty$. In fact we are interested in knowing how (1.26) behaves after the lapse of large time, i.e., when $t \gg t_c$, where t_c is some characteristic time, like period P, associated with the wave. The simplest method for obtaining this asymptotic value is the method of *steepest descent* or the *saddle-point method* because it demands the least possible details. (In Appendix I at the end of this chapter, we have briefly described this method. See also Jeffreys and Jeffreys (1946), and Dennery and Krzywicki (1967).)

We write (1.26) in the form

$$\phi(x,t) = \int_{-\infty}^{\infty} A(k)\exp\{it\chi(k)\}dk, \tag{1.28}$$

where the phase function $\chi(k)$ is given by

$$\chi(k) = \frac{x}{t}k - \omega(k). \tag{1.29}$$

We assume that $\chi(k)$ is analytic in the complex k-plane for a fixed value of x/t. In most of the physical problems of interest, this assumption is always valid.

A saddle point is defined as a point where the phase function $\chi(k)$ attains a stationary value. Therefore, in the present case, the saddle points are given by

$$\left.\frac{\partial\chi(k)}{\partial k}\right|_{(x/t)\text{ fixed}} = 0, \tag{1.30a}$$

i.e. by

$$\omega'(k) = x/t, \text{ provided } \omega''(k) \neq 0. \tag{1.30b}$$

On solving for k, we get the saddle points

$$k_i = k_i(x/t). \tag{1.30c}$$

Since the path of integration is along the real line, it is sufficient to consider only real saddle points k_i.

Corresponding to the saddle point k_i, the saddle-point methods gives the following asymptotic value for $\phi(x,t)$

$$\phi(x,t) \simeq \frac{\sqrt{(2\pi)}\,A(k_i)\exp\{it\,\chi(k_i) + i\alpha\}}{\{t|\chi''(k_i)|\}^{1/2}} \tag{1.31a}$$

$$= \frac{\sqrt{(2\pi)}\,A(k_i)\exp[i\{k_ix - \omega(k_i)t\} + i\alpha]}{\{t|\omega''(k_i)|\}^{1/2}} \tag{1.31b}$$

$$\text{as } t \to \infty,$$

where

$$\alpha = \frac{\pi}{4}, \text{ if } \omega''(k_i) < 0, \text{ i. e. if } \chi \text{ has a minimum value at } k_i$$

$$= -\frac{\pi}{4}, \text{ if } \omega''(k_i) > 0, \text{ i.e. if } \chi \text{ has a maximum value at } k_i. \tag{1.32}$$

Every other saddle point contributes to the value of $\phi(x,t)$ similarly and thus, taking all the saddle points, m in number, into account, we have

$$\phi(x,t) \simeq \sum_{i=1}^{m} \frac{A(k_i)\sqrt{(2\pi)}\exp\left[i\{k_ix - \omega(k_i)t\} - \frac{i\pi}{4}\mathrm{sgn}\,\omega''(k_i)\right]}{\{t|\omega''(k_i)|\}^{1/2}}, \quad (1.33)$$

where

$$\begin{array}{rl} \mathrm{sgn}\,\omega''(k_i) = & -1, \quad \text{if } \omega''(k_i) < 0 \\ = & 1, \quad \text{if } \omega''(k_i) > 0 \end{array} \Bigg\}. \quad (1.34)$$

The asymptotic expression (1.31b) for $\phi(x,t)$ appears surprising in many ways:

(i) it represents a locally harmonic wave which is not uniform in the sense that, in view of (1.30c), k_i and $\omega(k_i)$ vary with x and t (through the combination x/t), in spite of the fact that the initial state of the wave was not harmonic;

(ii) ultimately, i.e. when $t \gg P$, a phase difference is introduced which is equal to $\dfrac{\pi}{4}$ if the group velocity $\omega'(k)$ decreases with k and is equal to $-\dfrac{\pi}{4}$ if the group velocity $\omega'(k)$ increases with k; and

(iii) when $\omega''(k_i) \neq 0$, the amplitude $\bar{A}(t)$ of the wave given by

$$\bar{A}(t) = \frac{\sqrt{(2\pi)}\,A(k_i)}{\{t|\omega''(k_i)|\}^{1/2}} \quad (1.35)$$

decreases inversely as the square root of t over distances and times of the order of x and t themselves as seen from the following discussion of the relative changes in k with x and t. Assuming that $\omega''(k_i) \neq 0$ and rewriting (1.30b), we have for a saddle point k

$$x = \omega'(k)t. \quad (1.36)$$

On differentiating partially with respect to x and t, we can easily show that

$$\frac{k_x}{k} = \frac{\omega'(k)}{k\omega''(k)}\frac{1}{x} = O\left(\frac{1}{x}\right) \quad (1.37)$$

and

$$\frac{k_t}{k} = -\frac{\omega'(k)}{k\omega''(k)}\frac{1}{t} = O\left(\frac{1}{t}\right). \quad (1.38)$$

Since we are considering large values of time $(t \gg P)$ and large distances $(x \gg \lambda)$, the above expressions predict relative changes of $O(1)$ only over times and distances of order $T(\sim t)$ and $L(\sim x)$ respectively. Therefore, over distances of order d and times of order τ, where

$$\lambda \ll d \ll L \text{ and } P \ll \tau \ll T,$$

the changes in k, and hence in $\omega(k)$, may be neglected. In fact, it is on this reasoning that we have ignored the changes in $A(k_i)$ and $\omega''(k_i)$ while describing the asymptotic behaviour of $\bar{A}$ with t on the basis of the expression (1.35). A powerful nonlinear theory of dispersive waves presented in Chapter 5 will be developed on this observation.

At first sight, the decay of amplitude of the wave in a non-dissipative system appears surprising, but it is quite clear that this is an outcome of the distribution of the energy of the initial wave over the longer and longer wave train that results from the ever increasing dispersion as time passes. That our argument is sound can be seen from the following elementary consideration. The energy between the wave numbers k_i and $k_i + \mathrm{d}k$ is initially proportional to $A^2(k_i)\mathrm{d}k$. After time t, the distance between these wave numbers becomes

$$|t\omega'(k_i) - t\omega'(k_i + \mathrm{d}k)| \simeq t|\omega''(k_i)|\mathrm{d}k$$

so that now the energy density $\propto \dfrac{A^2(k_i)\mathrm{d}k}{t|\omega''(k_i)|\mathrm{d}k}$. Since the energy density of the wave is proportional to the square of the amplitude, the amplitude of the wave at t is proportional to $\dfrac{A(k_i)}{\{t|\omega''(k_i)|\}^{1/2}}$.

N.B. We note that when $\omega''(k_i) = 0$, the above discussion has to be significantly modified. Lighthill (1965), using the theory of the asymptotic behaviour of Fourier integrals has shown that the contribution of the saddle point k_i to $\phi(x,t)$ is asymptotically given by

$$\phi(x,t) \simeq \frac{A(k_i)\exp\left[i\{k_i x - \omega(k_i)t\}\right]\left(\tfrac{1}{3}\right)!\sqrt{3}}{\left[\tfrac{1}{6}t|\omega'''(k_i)|\right]^{1/3}} \qquad (1.39)$$

when $\omega'''(k_i) \neq 0$, so that the amplitude now falls of as inverse cube root of t.

1.6. Propagation of energy in a dispersive wave

We shall now determine the velocity with which the energy is propagated in a dispersive wave.

Let us consider two waves with wave numbers k_1 and k_2 starting from the point $x = 0$ at $t = 0$ with velocities $V_1(k_1)$ and $V_2(k_2)$ respectively. After sufficiently large time t, these waves will be at x_1 and x_2 where

$$x_1 = V_1(k_1)t \text{ and } x_2 = V_2(k_2)t \qquad (1.40)$$

provided V_1 and V_2 are the group velocities corresponding to wave numbers k_1 and k_2 respectively. At time t the value of ϕ is given by (1.31b) which represents *approximately* an harmonic wave if we neglect the *slow* variations in k_i and $\omega(k_i)$ with x and t. Therefore, the energy E of the wave between x_1 and x_2 may be taken as proportional to

$$\int_{V_1 t}^{V_2 t} \frac{2\pi A^2(k_i)\sin^2\left\{ k_i x - \omega(k_i)t \pm \dfrac{\pi}{4} \right\}dx}{t|\omega''(k_i)|}. \tag{1.41}$$

If we consider values of t such that $P \ll t \ll T$, then there will be several waves between x_1 and x_2 and we can treat k_i and $\omega(k_i)$ approximately constant. Taking the average of the harmonic term inside the integral in (1.41), we have

$$E \propto \int_{V_1 t}^{V_2 t} \frac{\pi A^2(k_i)dx}{t|\omega''(k_i)|}, \tag{1.42}$$

which, on substituting $x = Vt$ reduces to

$$E \propto \int_{V_1}^{V_2} \frac{\pi A^2(k_i)dV}{|\omega''(k_i)|}. \tag{1.43}$$

But from (1.30b) and (1.30c), we have

$$k_i = k_i(V), \quad V = \omega'(k_i) \tag{1.44}$$

and we may treat $A(k_i)$ and $\omega''(k_i)$ as functions of V so that the integration can be performed in principle and the result will depend only on V_1 and V_2 without having any reference to x and t. Consequently, we conclude that the energy between two points of the wave starting at the origin and moving with constant speeds is independent of t provided these speeds are the local group velocities.

From eqn (1.40), we have

$$x_2 - x_1 = [V_2(k_2) - V_1(k_1)]t, \tag{1.45}$$

so that $x_2 - x_1$ increases linearly with t. Our derivation of the asymptotic form (1.31b) shows that a given wave number is found at points given by (1.30b). This means that if an observer moves with the group velocity $\omega'(k)$, he will always be moving with a wave of wave number k. We shall give another proof of this in the next section.

We note the following important point. If we wish to follow a wave of a given wave number, we must move with velocity $V_g(k)$. Thus, after time $t(\gg P)$, this wave will be at

$$x_1 = V_g(k)t \tag{1.46}$$

and not at

$$x_2 = V_p(k)t. \tag{1.47}$$

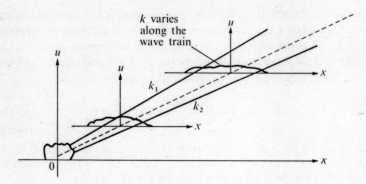

FIG. 1.3. Dispersion of waves starting at $x = 0$, $t = 0$.

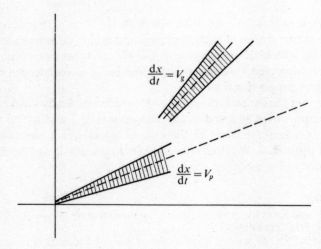

FIG. 1.4. Switching of wave moving with phase velocity to group velocity.

At x_2, there will be a wave of some other wave number k' such that $V_g(k') = V_p(k)$.

Moreover, a wave initially is propagated with velocity $V_p = \omega/k$ and after the lapse of sufficiently large time $t(\gg P)$, it is propagated with the group velocity $V_g = \omega'(k)$.

Figs. 1.3 and 1.4 summarize the above facts.

1.7.   An important kinematical relation

Let us consider a unit length in a fully dispersed progressive periodic wave in which the wave number and frequency vary with time and space. From the definition of wave number it is clear that, at an instant t, there will be k waves or k wave-crests in this length, so that k_t gives the rate of change of wave crests per unit time in the unit length. From the definition of the wave frequency, the number of waves or wave crests passing through a fixed point per unit time is ω, so that ω_x gives the net flux of waves across its two ends of the unit length. If we assume that the wave crests are neither created nor annihilated, we have the following extremely important kinematical relation

$$k_t + \omega_x(k) = 0, \text{ or } \frac{\partial k}{\partial t} + \omega'(k)\frac{\partial k}{\partial x} = 0 \tag{1.48}$$

in view of the dispersion relation. Thus along the characteristic $\dfrac{\mathrm{d}x}{\mathrm{d}t} = \omega'(k)$, $k = $ constant. Therefore, an observer moving with velocity

$$\frac{\mathrm{d}x}{\mathrm{d}t} = \omega'(k) \equiv V_g(k) \tag{1.49}$$

will always be with a wave of wave number k.

Since in our treatment of energy propagation in §1.6, the waves are assumed to be moving with local group velocity, the wave number (wavelength) along each wave is conserved. Consequently, the number of waves between x_1 and x_2 also increases proportional to t.

We note that the conservation law (1.48) for k is identically satisfied in case of the uniform waves as k and ω are independent of x and t. This relation, therefore, becomes important in the case of waves which have undergone significant dispersion. We shall refer to (1.48) frequently in our discussion of group velocity of a nonlinear wave.

Bibliography

DENNERY, P. and KRZYWICKI, A. *Methods of mathematical physics*, pp. 87–93. Harper and Row, New York (1967).

EDDINGTON, A. S. *The internal constitution of stars*. Cambridge University Press (1926).

JEFFREYS, H. and JEFFREYS, B. S. *Methods of mathematical physics*, §17.04 and 17.05. Cambridge University Press (1956).

LIGHTHILL, M. J. Group velocity. *J. Ins. Math. and its Appl.* **1**, 1–28 (1965).

APPENDIX I

Saddle-point method

The method of 'steepest descent' also called Debye's 'saddle point' method is given in almost all important books on methods of mathematical physics. We shall explain the salient points of this method here for ready reference.

Let us consider the integral of the type

$$I(t) = \int_c \exp\{tf(z)\}g(z)dz, \tag{I.1}$$

where, without loss of generality, we take t real and positive and assume the functions $f(z)$ and $g(z)$ as analytic in a certain region of the complex z-plane containing the contour c of integration. We are interested in the asymptotic value of $I(t)$ as $t \to \infty$.

We set $f(z) = \phi(x, y) + i\psi(x, y)$, where ϕ and ψ are real functions of $x = \mathrm{Re}\, z$ and $y = \mathrm{Im}\, z$. We then write the exponential function in the integrand as

$$\exp\{tf(z)\} = \exp(t\phi).\exp(it\psi). \tag{I.2}$$

The contribution of the exponential factor of the integrand to $I(t)$ comes from the regions of c where ϕ attains a relative maximum value (clearly ϕ and ψ cannot attain absolute maxima or absolute minima, $f(z)$ being analytic) provided the oscillations produced in the value of the exponential factor by $\exp(it\psi)$ do not offset the contribution of $\exp(t\phi)$. Clearly, therefore, we must first find the points on c where $\phi(x, y)$ attains relative maxima and deform the contour c in the neighbourhood of each of these points in such a manner that along these deformed paths ψ is constant, to eliminate the oscillations in the value of the exponential factor.

The stationary points of $f(z)$ are given by

$$f'(z) = 0. \tag{I.3}$$

Let z_0 be one of the roots of (I.3). Let c_0 be the deformed path passing through z_0 such that $\phi = \mathrm{Re}\, f(z)$ attains a relative maximum value at z_0 and $\psi = \mathrm{Im}\, f(z)$ is constant along c_0. Thus, if z is a point in the neighbourhood of z_0, we choose c_0 such that

$$\mathrm{Im}\, f(z) = \mathrm{Im}\, f(z_0) \tag{I.4}$$

as mentioned above.

By Taylor's theorem, we can expand $f(z)$ in the neighbourhood of z_0:

$$f(z) = f(z_0) + \frac{1}{2!}(z - z_0)^2 f''(z_0) + \ldots$$

since $f'(z_0) = 0$ by eqn (I.3).

By taking z sufficiently close to z_0, we can approximately set, without committing any significant error,

$$f(z) = f(z_0) + \frac{1}{2!}(z - z_0)^2 f''(z_0),\qquad\text{(I.5)}$$

which, in view of eqn (I.4), reduces to

$$\text{Re}\,[\,f(z) - f(z_0)] = \tfrac{1}{2}(z - z_0)^2 f''(z_0).\qquad\text{(I.6)}$$

Therefore, the right-hand side is also real.

Let us now set in eqn (I.5)

$$z - z_0 = r \exp(i\theta)\qquad\text{(I.7)}$$

and

$$\tfrac{1}{2}f''(z_0) = R \exp(i\alpha),\qquad\text{(I.8)}$$

where, z_0 is a fixed point, and R and α are constants. Thus, we have

$$\text{Re}\,[\,f(z) - f(z_0)] = r^2 R \cos(2\theta + \alpha)\qquad\text{(I.9)}$$

and

$$\text{Im}\,[\,f(z) - f(z_0)] = r^2 R \sin(2\theta + \alpha).\qquad\text{(I.10)}$$

In view of eqn (I.4), we have along c_0

$$\sin(2\theta + \alpha) = 0,\ \text{or}\ \theta = -\frac{\alpha}{2} + \frac{n\pi}{2},\ n = 0, 1, 2, 3.\qquad\text{(I.11)}$$

Substituting this value of θ in (I.7) we have

$$z = z_0 \pm r \exp(-i\alpha/2) \left.\begin{array}{l} + \ \text{for}\ n = 0 \\ - \ \text{for}\ n = 2 \end{array}\right\}\qquad\text{(I.12)}$$

and

$$z = z_0 \pm r \exp\{i(-\alpha/2 + \pi/2)\} \left.\begin{array}{l} + \ \text{for}\ n = 1 \\ - \ \text{for}\ n = 3. \end{array}\right\}\qquad\text{(I.13)}$$

Eqns (I.12) and (I.13) represent, in the z-plane, two straight lines passing through z_0 and inclined at angles $-\alpha/2$ and $\pi/2 - \alpha/2$ to the Re z-axis.

Similarly, $\qquad\qquad\qquad \text{Re}\,[\,f(z) - f(z_0)] = 0,$

when $\qquad\qquad\qquad\qquad \cos(2\theta + \alpha) = 0$

or when $\qquad\qquad\qquad \theta = (2n + 1)\dfrac{\pi}{4} - \dfrac{\alpha}{2},\ n = 0, 1, 2, 3.$

Substituting these values in (I.7), we obtain two straight lines passing through z_0 and inclined at angles $\pi/4 - \alpha/2$ and $3\pi/4 - \alpha/2$ to Re z-axis:

$$z = z_0 \pm r \exp\{i(\pi/4 - \alpha/2)\} \left.\begin{array}{l} + \ \text{for}\ n = 0 \\ - \ \text{for}\ n = 2 \end{array}\right\}\qquad\text{(I.14)}$$

and

$$z = z_0 \pm r \exp \{i(3\pi/4 - \alpha/2)\} \left. \begin{array}{l} + \text{ for } n = 1 \\ - \text{ for } n = 3. \end{array} \right\} \tag{I.15}$$

The straight lines (I.14) and (I.15) divide the z-plane in the neighbourhood of z_0 into four sectors, in which alternately

$$\text{Re } f(z) > \text{Re } f(z_0)$$

and

$$\text{Re } f(z) < \text{Re } f(z_0)$$

as shown in Fig. I.1.

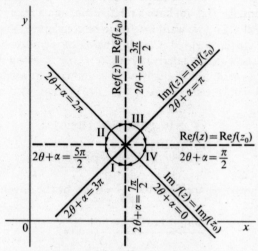

FIG. I.1. Division of z-plane near z_0 into four regions in which Re $f(z) > $ Re $f(z_0)$ and Re $f(z)$ < Re $f(z_0)$. Figure also shows lines along which Im $f(z)$ is constant.

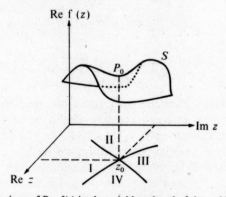

FIG. I.2. Behaviour of Re $f(z)$ in the neighbourhood of the saddle point z_0.

Since between two consecutive zeros of a cosine function, there is one and only one zero of the corresponding sine function, in each sector there is one and only one line on which Im $f(z)$ is constant. We have also marked these straight lines in Fig. I.1. In Fig. I.2, we have drawn the surface S traced by Re $f(z)$ against Re z and Im z, where P_0 on S corresponds to the point z_0. The points on S corresponding to the points z situated in sectors, say II and IV, are situated lower than P_0, while the points on S corresponding to the points in sectors I and III are situated higher than P_0. Thus, we can visualize S having the shape of horse's saddle (whence the method derives its name: the saddle-point method).

Now the choice of the path c_0 through z_0 along which integration should be performed to ensure the largest contribution to $I(t)$ by as small a part of c_0 as possible is evident: Re $f(z)$ will have a relative maxima at z_0 if c_0 lies in sectors II and IV. Since along c_0, we want Im $f(z)$ to be constant, c_0 must lie along the lines on which $\sin(2\theta + \alpha) = 0$. Clearly, therefore, c_0 must be taken along $A_2 A_4$. We now show that the derivative of Re $f(z)$ at z_0 along the line $A_2 A_4$ is the largest. Let us write the derivatives of ϕ and ψ at z_0 along the direction making angle θ with the Re z-axis,

$$\phi_s = (\phi_x) \cos\theta + (\phi_y) \sin\theta$$

and

$$\psi_s = (\psi_x) \cos\theta + (\psi_y) \sin\theta.$$

The extreme values of ϕ_s are given by $\dfrac{\partial \phi_s}{\partial \theta} = 0$, i.e. by the directions θ given by

$$\frac{\sin\theta}{\phi_y} = \frac{\cos\theta}{\phi_x} = \frac{1}{\sqrt{(\phi_x^2 + \phi_y^2)}} \tag{I.16}$$

and then

$$\frac{\partial^2 \phi_s}{\partial \theta^2} = -\frac{2\phi_x \phi_y}{\sqrt{(\phi_x^2 + \phi_y^2)}}. \tag{I.17}$$

In this direction

$$\psi_s = \frac{\psi_x \phi_x + \psi_y \phi_y}{\sqrt{(\phi_x^2 + \phi_y^2)}} = 0 \tag{I.18}$$

in view of the Cauchy–Riemann equations. This completes the proof of the desired result, namely ϕ_s attains maximum value in the direction θ in which the Im $f(z)$ is constant. Thus along $A_2 A_4$ the slope of Re $f(z)$ is as steep as possible. From this property, the method derives its name 'method of steepest descent'.

We can easily extend the foregoing discussion to take account of higher order terms in (I.5), but now the paths along which Im $f(z)$ is constant are curves having lines $A_1 A_3$ and $A_2 A_4$ as tangents at the saddle point z_0.

Rewriting (I.5) as

$$f(z) = f(z_0) - \xi^2, \qquad (I.19)$$

where

$$\xi^2 = -\tfrac{1}{2}(z - z_0)^2 f''(z_0), \qquad (I.20)$$

we find that ξ^2 is real positive, as (I.19) essentially means

$$\text{Re } f(z_0) - \text{Re } f(z) = \xi^2,$$

as the imaginary part is constant along c_0 and at z_0 there is a relative maximum of Re $f(z)$. The real variable ξ takes negative and positive values as z moves on c_0 in the neighbourhood of z_0.

Setting for $z - z_0$ from (I.7) and for $\tfrac{1}{2} f''(z_0)$ from (I.8) in (I.20), we get

$$\xi^2 = - Rr^2 \exp\{i(2\theta + \alpha)\} > 0$$

so that

$$\left. \begin{array}{c} \xi = R^{1/2} r \text{ on one side of } z_0 \text{ and } \xi = - R^{1/2} r \\ \text{on the other side } z_0 \text{ on the path } c_0. \end{array} \right\} \qquad (I.21)$$

Therefore, from eqns (I.7) and (I.21), we have

$$z - z_0 = \frac{\pm \, \xi \exp(i\theta)}{R^{1/2}} \qquad (I.22)$$

or

$$\frac{dz}{d\xi} = \pm \frac{\sqrt{2} \exp(i\theta)}{|f''(z_0)|^{1/2}}, \qquad (I.23)$$

where we have to take only one of the two signs for all points on c_0 (see the comment after the equation (I.26)).

Substituting for $f(z)$ from (I.19) in (I.1), we have

$$I_0(t) \simeq \exp\{tf(z_0)\} \int_{c_0} \exp\{- t\xi^2(z)\} g(z) dz, \qquad (I.24)$$

where c_0 is the path of steepest descent. When t is positive and large, the dominant contribution to $I_0(t)$ comes from the regions along c_0 where ξ is small; for when ξ is large, the integrand falls exponentially in its value. Therefore, we shall be *making only a negligible error*, if we replace the integral over a region around $\xi = 0, |\xi| \ll 1$ by an integral with respect to ξ extending from $\xi = -\infty$ to $\xi = \infty$.

Thus, we have

$$I_0(t) \simeq \exp\{tf(z_0)\} \int_{-\infty}^{\infty} \exp(- t\xi^2) g(z) \frac{dz}{d\xi} d\xi$$

$$\simeq \frac{\pm \sqrt{2} \exp(i\theta) \exp\{tf(z_0)\}}{|f''(z_0)|^{1/2}} \int_{-\infty}^{\infty} \exp(- t\xi^2)\{g(z_0)$$

$$+ (z - z_0)g'(z_0) + \frac{1}{2!}(z - z_0)^2 g''(z_0) + \ldots\} \, dz$$

$$\simeq \frac{\pm\sqrt{2} \exp{(i\theta)} \exp{\{t f(z_0)\}}}{|f''(z_0)|^{1/2}} \left[g(z_0)\sqrt{\frac{\pi}{t}} + \frac{\exp{(2i\theta)}g''(z_0)}{|f''(z_0)|} \frac{\sqrt{\pi}}{2t^{3/2}} + \ldots \right]$$

$$\text{(I.25)}$$

so that when t is large

$$I_0(t) \simeq \frac{\pm\sqrt{(2\pi)} \exp{(i\theta)} \exp{\{t f(z_0)\}}}{|t f''(z_0)|^{1/2}} g(z_0). \qquad \text{(I.26)}$$

In the range $(-\pi, \pi)$ there are two possible choices for θ and they differ by π. In any practical application we have to decide from the behaviour of real and imaginary parts of $f(z)$ in what sense the path of integration passes through the saddle point. If we select the value of θ that makes r positive at points on the path after passing through z_0, we shall have to take the positive sign in (I.22) as ξ goes from $-\infty$ to ∞ on the path.

We note that the asymptotic expansion in (I.25) has to be understood in the Poincare sense, namely we say that $\sum\limits_{k=0}^{\infty} c_k/z^k$ is an asymptotic expansion for $f(z)$, if for any positive integer n,

$$\lim_{|z| \to \infty} \left[z^n\left\{ f(z) - \sum_{k=0}^{n} c_k/z^k \right\} \right] = 0,$$

irrespective of the fact whether the series $\sum\limits_{k=0}^{\infty} c_k/z^k$ converges or not. The sum of a finite number of terms of the series may represent $f(z)$ quite accurately if $|z|$ is large, for

$$\left| f(z) - \sum_{k=0}^{n} c_k/z^k \right| = O\left\{ \frac{1}{|z|^{n+1}} \right\} \text{ as } |z| \to \infty.$$

2

SOME NONLINEAR EQUATIONS
OF EVOLUTION
(Steady Solution)

2.1. Introduction

It is well known that the initial and boundary value problems associated with nonlinear partial differential equations are very difficult to handle in a general way. Some specific problems have been tackled from time to time by methods specifically suited to the individual problems. Nonlinear waves, in which we are now interested, are governed by nonlinear partial differential equations. In this chapter we shall concentrate on some simple nonlinear wave model equations which have received considerable attention during the past decade or so. This will help us in clearly bringing out, with comparative ease, the role of such factors as nonlinearity, diffusion, and dispersion in the spatial and temporal evolution of an entity. In fact, this is the main objective of this monograph. In consonance with the above objective, we shall make a comparative study of the following two sets; Set I consisting of linear equations and Set II consisting of nonlinear equations

Set I *linear*
- (a) $u_t + cu_x = 0$, $c = $ constant;
- (b) $u_t + cu_x - \mu u_{xx} = 0$, c, μ constant, $\mu > 0$; and
- (c) $u_t + cu_x + K u_{xxx} = 0$, c, K constant, $K > 0$.

Set II *nonlinear*
- (a) $u_t + uu_x = 0$;
- (b) $u_t + uu_x - \mu u_{xx} = 0$ (Burgers equation); and
- (c) $u_t + uu_x + K u_{xxx} = 0$ (Korteweg–de Vries (KdV) equation).

The Burgers equation is the simplest model of diffusive waves and, under certain simplifying assumptions, covers the following cases amongst others: turbulence (where this equation had its origin), sound waves in viscous media, waves in fluid-filled visco-elastic tubes, and magnetohydrodynamic waves in media having finite electrical conductivity. The KdV equation is the simplest model of the dispersive waves and, under certain simplifying conditions, covers cases of the following type: surface waves of long wavelength in liquids, plasma waves, lattice waves, weakly nonlinear magnetohydrodynamic waves. The wide applicability of these equations is the main reason why, during the past decade or so, they have attracted so much attention from mathematicians.

We have defined the diffusive and dispersive waves in Chapter 1 with the help of the dispersion relation obtained by the Fourier method. We cannot apply the Fourier method to the nonlinear equations and therefore we must find some other way of classifying these waves. It is customary to say that the wave represented by a nonlinear equation is diffusive or dispersive according as the wave represented by the corresponding linearized equation is diffusive or dispersive. In the present chapter our effort will be to determine the comparative roles of the nonlinear terms in these equations and the terms containing the spatial derivatives of order higher than the first.

2.2. Effect of nonlinearity

To study the effect of nonlinearity, we solve I(a) and II(a) in §2.1 under the same initial conditions. The initial value is intentionally chosen to be simple so that the physical facts are not lost in the complicated mathematical expressions.

2.2.1 Solution of I(a)

This equation is linear and using the method of Lagrange we can immediately write its general solution as

$$u(x,t) = f(x - ct),\qquad(2.1)$$

where f is an arbitrary function. Eqn (2.1) represents a wave moving with velocity c in the positive direction of the x-axis.

Let us now particularize f by assuming the following initial condition:

$$\left.\begin{aligned} u(x,0) &= a^2 - x^2, \text{ for } |x| \leq a \\ &= 0, \qquad\quad \text{ for } |x| > a, \end{aligned}\right\} \qquad(2.2)$$

where $a > 0$. Eqn (2.2) represents a parabolic pulse spread over $-a \leq x \leq a$.

In view of eqn (2.2), the solution (2.1) becomes

$$\left.\begin{aligned} u(x,t) &= a^2 - (x - ct)^2, \text{ for } |x - ct| \leq a \\ &= 0 \qquad\qquad , \text{ for } |x - ct| > a \end{aligned}\right\} \qquad(2.3)$$

In terms of the moving coordinate defined by the relation

$$\xi = x - ct \qquad(2.4)$$

the above solution can be written in the following convenient form

$$\left.\begin{aligned} u(x,t) &= a^2 - \xi^2, \text{ for } |\xi| \leq a \\ &= 0 \qquad , \text{ for } |\xi| > a. \end{aligned}\right\} \qquad(2.5)$$

Eqn (2.5) does not have any *explicit* dependence on time t, hence it is called *steady*. In the present chapter we shall mostly consider only such steady solutions.

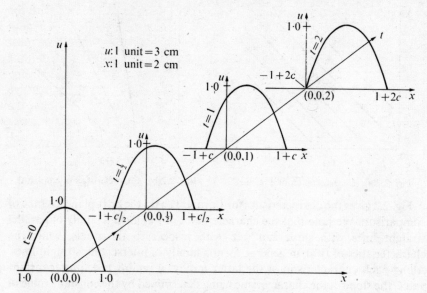

FIG. 2.1. Development of a parabolic pulse according to eqn. I(a) for $a = 1$, $c = 1/2$.

Fig. (2.1) represents the propagation (or evolution) of the initial parabolic pulse with x and t.

The solution (2.5) represents the same pulse as the initial one whose centre has shifted by ct in the positive direction of the x-axis during time t. We describe this fact by saying that a solutions of I(a) represents a wave moving with velocity c in the positive direction of the x-axis *without change of form*.

2.2.2   Solution II(a)

To study the effect of nonlinearity, we now solve II(a) under the same initial condition as given in eqn (2.2).

The characteristic equations for II(a) are

$$\frac{dt}{1} = \frac{dx}{u} = \frac{du}{0} \tag{2.6}$$

so that along the characteristic

$$\frac{dx}{dt} = u \tag{2.7}$$

u is conserved and the general solution of II(a) is

$$u(x,t) = f(x - ut) \tag{2.8}$$

where f is an arbitrary function.

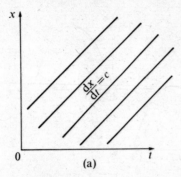

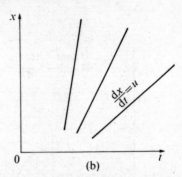

Fig. 2.2(a). Characteristics of eqn. I(a)   Fig. 2.2(b). Characteristics of eqn. II(a)

Fig. 2.2 gives the characteristics for I(a) and II(a) in the (x,t)-plane for sake of comparison. We note that the characteristics of I(a) form a family of parallel straight lines with slope $\tan^{-1} c$ with respect to the t-axis, while the characteristics of II(a), in general, form a family of intersecting straight lines. Along each characteristic of the latter family, u retains a definite constant value, the slope of the characteristic being determined by the constant value of u on it.

The basic idea of wave propagation governed by a hyperbolic equation can be briefly described as a property associated with a disturbance which moves with a finite velocity. In this sense, we can treat each characteristic in the (x,t)-plane as a moving wavelet, and the property of the wave that remains constant along an individual characteristic as the bit of information that it carries with it. In this sense, I(a) represents a system of wavelets all moving with the same constant velocity c and the constant value of u associated with a characteristic represents the bit of information that it carries with it. Similarly, II(a) represents a system of wavelets each moving with a different velocity. The wavelet which carries a higher value of u moves faster.

In order to study the effect of nonlinearity on the wave profile as it is propagated, we shall obtain the solution of II(a) satisfying the initial condition (2.2). Under the initial condition (2.2), the solution (2.8) becomes

$$\left.\begin{array}{ll} u(x,t) = a^2 - \xi^2, & \text{for } |\xi| \leq a \\ \quad = 0, & \text{for } |\xi| \geq a, \end{array}\right\} \tag{2.9}$$

where

$$\xi = x - ut. \tag{2.10}$$

On solving (2.9) explicitly for u, we have

$$\left.\begin{array}{ll} u(x,t) = \dfrac{1}{2t^2}[(2xt - 1) \pm (1 - 4xt + 4a^2t^2)^{1/2}] & \\ \quad = 0, & \text{for } |x - ut| \leq a \\ & \text{for } |x - ut| > a. \end{array}\right\} \tag{2.11}$$

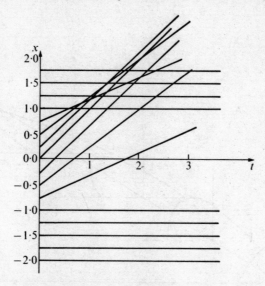

Fig. 2.3 Characteristics of eqn. II(a) with initial condition (2.2), $a = 1$.

When t is small ($t \to 0$), only the upper sign before the radical in (2.11) is admissible, in order that the initial condition is satisfied. When $t > T$ (to be determined), both the signs in (2.11) are admissible for $x > a$.

Fig. 2.3 depicts the characteristics of II(a) with initial condition (2.2) in the (x, t)-plane with $a = 1$. Fig. 2.3 is very instructive as it indicates graphically the propagation of wavelets starting from the various points x on the x-axis at time $t = 0$. All wavelets starting from points $(x, 0)$ where $x > -1$ intersect the characteristics starting from the points x where $x \geq 1$ at some time or another. This intersection of characteristics results in the association of more than one value of u at the point of intersection. Clearly, this sort of situation is physically *untenable*. Consequently, if we are interested in a unique bounded solution in such a situation, we have to perforce introduce the concept of *weak solution* which permits moving jump discontinuities. These discontinuities are called *shocks* in fluid dynamics.

From Fig. 2.3, it is also evident that the points $x = \pm a$ remains fixed for all times.

Fig. 2.4 describes the propagation of the pulse (2.2) taking $a = 1$. As t increases, the u-profile continually gets progressively more and more deformed. From this we conclude that the nonlinearity brings about a progressive deformation of the initial wave profile.

We shall now determine T in this special case. The initial profile has both positive and negative slopes. Specifically, at $x = a$, the slope is negative

$$u_x(a, 0) = -2a < 0. \tag{2.12}$$

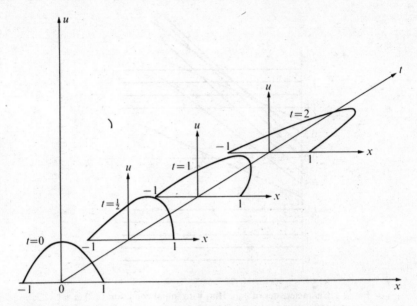

FIG. 2.4. Propagation of a parabolic pulse according to eqn. II(a) for $a = 1$.

Clearly, $u(x,t)$ can have multiple values for $x \geq a$ only when $u_x(a,t) > 0$. The minimum value of t for which this inequality holds is our T. Also, $u_x(a,t)$ can change from negative to positive only through infinity. Now, from (2.11), we have

$$u_x(a,t) = \frac{1}{2t^2}\left[2t \mp \frac{2t}{1 - 2at} \right];$$ (2.13)

therefore,

$$T = \frac{1}{2a}.$$ (2.14)

We can easily extend the foregoing discussion to a more general pulse-profile than that assumed in (2.2). Let the initial profile be given by

$$\left. \begin{aligned} u(x,0) &= f(x), \text{ for } |x| \leq a \\ &= 0, \quad \text{ for } |x| > a. \end{aligned} \right\}$$ (2.15)

where $f(x)$ is a continuously differentiable function with positive and negative slopes.

The solution of II(a) under the condition (2.15) is given by

$$\left. \begin{aligned} u(x,t) &= f(\xi), \xi = x - ut, \text{ for } |\xi| \leq a \\ &= 0, \quad \text{ for } |\xi| > a. \end{aligned} \right\}$$ (2.16)

where ξ is the spatial coordinate moving with velocity u (which is itself a function of x and t). The role of ξ is similar to the Lagrangian coordinate in Lagrangian formulation of fluid dynamical equations. At $t = 0$, $x = \xi$.

Differentiating (2.16) with respect to x partially, we get

$$u_x(x,t) = f_\xi(1 - u_x t)$$

so that

$$u_x(x,t) = \frac{f_\xi}{1 + f_\xi t}, \tag{2.17}$$

which expresses the slope of the u-profile at the point (x,t) in terms of the slope of the initial profile at ξ, where x is the position at t of the point which was initially at ξ. If $f_\xi < 0$, $u_x(x,t)$ is infinite at $t = \left(\dfrac{1}{-f_\xi}\right)$. Therefore, if the initial profile has a negative slope at some point ξ, then for $t > T = \left(\dfrac{1}{-f_\xi}\right)_{\text{min.}}$, the solution ceases to be single valued in the neighbourhood of a point $x_0 = \xi_0 + Tf(\xi_0)$ where ξ_0 is the point at which $\left(\dfrac{1}{-f_\xi}\right)$ attains the minimum value.

Let us try to find out the changes in the slope of the pulse profile at a wavelet designated by $\xi = \xi_0$ when t crosses T. Let $X_0(t)$ be the position of the wavelet at any time t. We wish to find $u_x(X_0(t),t)$ for the values of

$$t = T + \varepsilon = \left(\frac{1}{-f_\xi}\right)_{\text{min.}} + \varepsilon$$

where $|\varepsilon|$ is small. Then

$$u_x(X_0(t), T + \varepsilon) = \left(\frac{f_\xi}{1 + f_\xi t}\right)_{\xi = \xi_0, t = T + \varepsilon}$$

$$= \frac{f_\xi(\xi_0)}{1 + f_\xi(\xi_0)\left[-\dfrac{1}{f_\xi(\xi_0)} + \varepsilon\right]}$$

$$= 1/\varepsilon.$$

Thus, we have

$$u_x(X_0(T - 0), T - 0) = -\infty \quad \text{and} \quad u_x(X_0(T + 0), T + 0) = +\infty$$

as mentioned earlier.

2.2.3 *Role of nonlinearity*

Therefore, we conclude that in the case of equations of the type I(a) and II(a) in §2.1

A linear wave travels without change of profile. The role of nonlinearity is to produce progressively more and more deformation in the wave profile as t increases. After some time $(t > T)$, a physically meaningful solution is the one which contains a moving jump discontinuity. We have called such a solution a weak solution.

We note that the linear equation I(a), when subjected to the Fourier method, yields the dispersion relation

$$\omega = ck$$

so that the phase velocity and the group velocity are both equal to c. In the case of the nonlinear equation II(a), we cannot apply the Fourier method and therefore we cannot talk about the wave number, frequency, wave velocity, group velocity unless and until we indicate the method of defining them. We shall take up this question in Chapter V.

2.3. Diffusive Waves

We shall now study the equations I(b) and II(b) in §2.1 to determine the effects produced by the presence of second order derivative u_{xx}. This comparative study will also indicate the effect of diffusion produced by this term on the deformation of the wave profile produced by the nonlinearity. In the context of fluid dynamics, the nonlinear term uu_x represents the convection term, while the second-order term represents the viscous force. Thus the present study will reveal competitive roles of the nonlinear convection in steepening (in 'compression regions' of the pulse) the profile and of viscous dissipation in broadening it.

2.3.1   Solution of I(b)

Setting

$$u(x,t) = a \exp\{i(kx - \omega t)\} \tag{2.18}$$

in I(b), we get the dispersion relation

$$\omega = ck - i\mu k^2. \tag{2.19}$$

Hence according to our definition introduced in Chapter 1, this equation represents a diffusive wave. Consequently, the wave represented by II(b) is also diffusive in accordance with the convention mentioned earlier in this chapter.

From (2.19), we have

$$\left.\begin{array}{l} \text{Re } \omega = ck \\ \text{Im } \omega = -\mu k^2 < 0, \end{array}\right\} \tag{2.20}$$

and

since we have chosen $\mu > 0$.

In view of (2.19), the wave profile (2.18) may be written as

$$u(x,t) = \{a\exp(-t/t_0)\}\exp\{ik(x-ct)\} \qquad (2.21)$$

which represents a harmonic wave with wave number k and velocity c whose amplitude decays (attenuates) exponentially with time. The decay time is

$$t_0 = \frac{1}{\mu k^2}. \qquad (2.22)$$

For a given μ, t_0 becomes smaller and smaller as k increases. Therefore, the waves of smaller lengths attenuate faster than waves of longer lengths. Thus, after a sufficiently long time ($t \gg t_0$) only the long wavelength disturbances will survive. Similarly, for a fixed k, t_0 decreases as μ increases; consequently the waves of a given wavelength attenuate faster in a medium with larger μ-value. In this sense we may regard μ as a measure of diffusion.

From eqn (2.21) we are able to define the phase velocity:

$$V_p = c \equiv \mathrm{Re}(\omega/k). \qquad (2.23)$$

2.3.2   Solution of II(b)

In the present chapter, we are interested in the steady solutions, therefore we shall seek the steady solution of the Burgers equation in the form

$$u(x,t) = u(\xi), \qquad \xi = x - ct. \qquad (2.24)$$

On substituting (2.24) in II(b), we have the following second-order ordinary differential equation:

$$-cu_\xi + uu_\xi - \mu u_{\xi\xi} = 0 \qquad (2.25)$$

which, on integration yields

$$-cu + \tfrac{1}{2}u^2 - \mu u_\xi = A \qquad (2.26)$$

where A is the *constant* of integration.

We now write (2.26) as

$$u_\xi = \frac{1}{2\mu}(u - u_\infty^-)(u - u_\infty^+), \qquad (2.27)$$

where

$$\left. \begin{array}{l} u_\infty^+ = c - \sqrt{(c^2 + 2A)} \\[4pt] u_\infty^- = c + \sqrt{(c^2 + 2A)} \end{array} \right\} \qquad (2.28)$$

and

are the two roots of

$$u^2 - 2cu - 2A = 0. \qquad (2.29)$$

The meanings of the subscript ∞ and the superscript $+$ and $-$ attached to u will be clear when we consider the behaviour of the solution as $\xi \to \pm\infty$.

In order to ensure that the roots $u_\infty^\pm$ are real, we shall assume that

$$c^2 + 2A > 0 \text{ and then } u_\infty^- > u_\infty^+. \tag{2.30}$$

On integrating (2.27), we have

$$u(x,t) = c - \sqrt{(c^2 + 2A)} \tanh\left(\frac{\sqrt{(c^2 + 2A)}}{2\mu}\xi\right) \tag{2.31}$$

$$= \frac{1}{2}\left((u_\infty^- + u_\infty^+) - (u_\infty^- - u_\infty^+)\tanh\left[\frac{u_\infty^- - u_\infty^+}{4\mu}\{x - \right.\right.$$

$$\left.\left. \tfrac{1}{2}(u_\infty^- + u_\infty^+)t\}\right]\right), \tag{2.32}$$

where we have chosen the constant of integration from the condition

which implies
$$\begin{aligned}\xi \to +\infty, u \to u_\infty^+; \\ \xi \to -\infty, u \to u_\infty^-.\end{aligned} \tag{2.33}$$

Thus the solution (2.31) joins the two asymptotic states u_∞^- at $\xi = -\infty$ and u_∞^+ at $\xi = +\infty$ through the continuously varying states. From (2.28), we have

$$c = \tfrac{1}{2}(u_\infty^- + u_\infty^+) \tag{2.34}$$

which evidently is the Rankine – Hugoniot relation for the present problem. Consequently, in the terminology of fluid dynamics, the solution (2.31) gives the structure of the shock wave.

For comparison, we record below the steady solution of the linearized form of the Burgers equation:

$$u_t - \mu u_{xx} = 0 \tag{2.35}$$

$$\left.\begin{aligned}u(x,t) = u(\xi) = A + B\exp\left(-\frac{c}{\mu}\xi\right), \\ \xi = x - ct \text{ with } c > 0, \text{ say}\end{aligned}\right\} \tag{2.36}$$

When
but when
$$\left.\begin{aligned}\xi \to +\infty \qquad u \to A, \\ \xi \to -\infty \qquad u \to \infty.\end{aligned}\right\} \tag{2.37}$$

Consequently the only bounded solution of (2.35) is a constant state. Thus the linearized form of Burgers equation does not admit a solution joining two uniform states through continuously varying states. Therefore, we conclude that

The nonlinearity of the Burgers equation achieves the smooth joining of two asymptotic uniform states through continuously varying states.

We have noticed earlier that if there is a region of negative slope in the pulse profile, the solution of II(a) (which does not contain the u_{xx} term) develops a very steep slope even when the initial profile does not have any steep slope. On the contrary it is not difficult to show that the presence of the second derivative term not only prevents the formation of very steep gradients but actually smoothens any initial discontinuity immediately (Lighthill (1956), section 7.3).

From this we conclude that the second order term in the Burgers equation does not permit the development of the steep slopes in the wave profile. Thus the second order term tends to counteract the effect of nonlinearity in the compression regions and tends to diffuse (spread) the sharp discontinuities into smooth profiles.

Let us denote by ξ^- the value of ξ where $u(\xi^-) = u_\infty^- - \alpha(u_\infty^- - u_\infty^+)$ and by ξ^+ the value of ξ where $u(\xi^+) = u_\infty^+ + \alpha(u_\infty^- - u_\infty^+)$, where α is a small positive fraction. Then, from (2.31), we have

$$\xi^+ - \xi^- = \frac{4\mu}{(u_\infty^- - u_\infty^+)}\{\ln(1-\alpha) - \ln\alpha\}. \qquad (2.38)$$

From eqn (2.38), the role of the parameter μ is clear. When $\mu \to 0, \xi^+ \to \xi^-$ so that there is a sharp discontinuity in $u(\xi)$ during transition from u_∞^- to u_∞^+. Thus μ tends to spread out the sharp discontinuity in u-profile which the nonlinearity tends to produce. It is in view of this property that we say that the wave is diffusive and μ is the measure of diffusion. In fact, (2.38) is a measure of the shock thickness in the fluid-dynamical terminology.

2.4.   Dispersive waves

We shall now discuss the equations I(c) and II(c) in § 2.1 to study the role of the term containing the third-order x-derivative in relation to nonlinearity.

For simplicity we consider the case $K > 0$. However, we note that though the nonlinear equation

$$u_t + uu_x + Ku_{xxx} = 0, \quad K < 0$$

can always be transformed to

$$u_t + uu_x + (-K)u_{xxx} = 0, \quad -K > 0$$

by the transformation:

$$u \to -u, \quad x \to -x, \quad t \to t,$$

we cannot simultaneously transform the linear equation to a form with $K > 0$.

2.4.1   *Solution of* I(c)

Applying the Fourier method to- the linear equation I(c), we get the following dispersion relation:

$$\omega = ck - Kk^3.$$

Thus ω is a real function of k and we can define the phase velocity and the group velocity in the usual manner:

$$V_p = \omega/k = c - Kk^2 \qquad \text{(a function of } k)$$

and

$$V_g = \omega'(k) = c - 3Kk^2 \qquad \text{(a function of } k)$$

so that $V_p \neq V_g$ and the wave is dispersive.

2.4.2. *Solution of the KdV equation*

Let us first consider the linearized form of the KdV equation

$$u_t + Ku_{xxx} = 0$$

for which the dispersion relation is

$$\omega = -Kk^3.$$

Since $\omega''(k) \neq 0$, the wave is dispersive.

We now seek the steady solution of the KdV equation in the form

$$u(x,t) = u(\xi), \qquad \xi = x - ct. \tag{2.39}$$

Substituting (2.39) in II(c) we have

$$-cu_\xi + uu_\xi + Ku_{\xi\xi\xi} = 0 \tag{2.40}$$

which on integration yields

$$-cu + \tfrac{1}{2}u^2 + Ku_{\xi\xi} = A. \tag{2.41}$$

Multiplying the last equation by u_ξ and integrating, we get

$$-\tfrac{1}{2}cu^2 + \tfrac{1}{6}u^3 + \tfrac{1}{2}Ku_\xi^2 = Au + B$$

or

$$3Ku_\xi^2 = -u^3 + 3cu^2 + 6Au + 6B \equiv f(u). \tag{2.42}$$

If we write it in the form

$$\tfrac{1}{2}u_\xi^2 + \frac{1}{6K}\{-f(u)\} = 0 \tag{2.43}$$

and interpret u and ξ as space and time coordinates respectively, we may interpret eqn (2.43) as the energy equation for the motion of a particle with unit mass moving under the action of the potential $\dfrac{1}{6K} f(u)$ or as an equation for an anharmonic oscillator. Under certain conditions eqn (2.43) may represent a

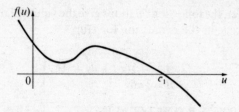

FIG. 2.5. A rough sketch of $f(u)$ when it has only one real zero.

periodic motion oscillating between two consecutive real zeros of $f(u)$ where $f(u) \geq 0$. This provides an important clue for studying the equation (2.42). Now $f(u)$ is cubic and therefore has three zeros. We first note that if $u = c$ is a zero of $f(u)$, then $u = c$ is a constant solution of (2.43). In what follows we shall study nonconstant bounded solutions. Evidently, we have to consider two cases: (i) when only one zero is real and (ii) when all the three zeros are real.

Case (i)   Fig. 2.5 gives a rough sketch of $f(u)$ against u when it has only one real zero at $u = c_1$.

From Fig. 2.5, it is clear that $f(u) < 0$ when $u > c_1$ and $f(u) \geq 0$ when $u \leq c_1$, so that, for a real solution, we must consider the range $u \leq c_1$. In this region we get

$$\frac{du}{d\xi} = \pm \left\{ \frac{1}{3K} f(u) \right\}^{1/2}.$$

Firstly, unless $f'(c_1) \neq 0$, the condition $u = c_1$ cannot be attained at a finite distance. Therefore, we assume $f'(c_1) \neq 0$. Now, a solution of the above equation exists satisfying $u = c_1$ at $\xi = \xi_0$. Since $f(u)$ remains positive for $u < c_1$ and tends to $+\infty$ as $u \to -\infty$, the solution becomes unbounded.

Case (ii)   We shall now consider the case when all the three zeros α, β, γ of $f(u)$ are real. Without loss of generally we may assume the following order of magnitude: $\alpha \geq \beta \geq \gamma$. Fig. 2.6 gives a rough sketch of $f(u)$ against u when all the roots are distinct (curve A), when $\beta = \gamma$ (curve B), and when $\beta = \alpha$ (Curve C). The curve B touches the u-axis at γ and the curve C touches the u-

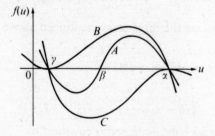

FIG. 2.6   A rough sketch of $f(u)$ against u when it has three real zeros. A: three distinct zeros (cnoidal wave); B: $\beta = \gamma$ (solitary wave); C: $\beta = a$ (constant solution $u \to a$).

axis at α. Moreover, the following relations give the values of the parameters C, A, and B that occur in the expression for $f(u)$

$$c = \tfrac{1}{3}(\alpha + \beta + \gamma),$$
$$A = -\tfrac{1}{6}(\alpha\beta + \beta\gamma + \gamma\alpha), \left. \right\} \tag{2.44}$$
and
$$B = \tfrac{1}{6}\alpha\beta\gamma$$

In terms of the roots α, β, γ we can write

$$f(u) = (u - \gamma)(u - \beta)(\alpha - u). \tag{2.45}$$

For the solution to be real and bounded, we have to restrict u between β and α for curve A, and between $\beta = \gamma$ and α for curve B. For curve C, the values of u $\leq \gamma$ are admissible but in this case as in Case (i) no bounded solution is possible. However, the discussion of the case when $\beta \to \alpha$ seems to be particularly interesting.

We shall first deduce a set of sufficient conditions under which $f(u)$ has three real zeros. We can easily check that

$$c^2 + 2A = \tfrac{1}{18}\{(\alpha - \beta)^2 + (\beta - \gamma)^2 + (\gamma - \alpha)^2\}$$

so that when α, β, γ are real,

$$c^2 + 2A \geq 0. \tag{2.46}$$

Now we note that $f(u)$ attains a positive maximum value between β and α and a negative minimum value between γ and β. The extreme values of $f(u)$ are attained at

$$u = c \pm \sqrt{(c^2 + 2A)}.$$

Therefore

$$f\{c + \sqrt{(c^2 + 2A)}\} \geq 0 \tag{2.47}$$

and

$$f\{c - \sqrt{(c^2 + 2A)}\} \leq 0. \tag{2.48}$$

Eqns (2.46)–(2.48) provide a required set of conditions for all the zeros of $f(u)$ to be real.

We shall now discuss the three cases mentioned above in turn.

Case A; α, β, γ, distinct.
From eqns (2.42) and (2.45), we have

$$-\frac{du}{[(u - \gamma)(u - \beta)(\alpha - u)]^{1/2}} = \frac{d\xi}{(3K)^{1/2}}. \tag{2.49}$$

Putting $\alpha - u = p^2$, we reduce the above equation to

$$\frac{d\xi}{(3K)^{1/2}} = \frac{2dp}{[\{(\alpha - \gamma) - p^2\}\{(\alpha - \beta) - p^2\}]^{1/2}}. \tag{2.50}$$

If we now substitute

$$p = \sqrt{(\alpha - \beta q)}$$

in the last equation, it takes the following familiar form

$$\sqrt{(\alpha - \gamma)} \frac{d\xi}{\sqrt{(3K)^{1/2}}} = \frac{2dp}{\{(1 - s^2 q^2)(1 - q^2)\}^{1/2}} \qquad (2.51)$$

where

$$s^2 = \frac{\alpha - \beta}{\alpha - \gamma}, \quad q = \frac{p}{\sqrt{(\alpha - \beta)}} = \sqrt{\left(\frac{\alpha - u}{\alpha - \beta}\right)}. \qquad (2.52)$$

From our ordering of the zeros, it is clear that $0 < s^2 < 1$.

As mentioned earlier $\beta \leq u \leq \alpha$. From (2.52) $u = \alpha$ corresponds to $q = 0$, while $u = \beta$ corresponds to $q = 1$. Therefore, choosing $\xi = 0$ at $q = 0$ we have

$$\xi = \sqrt{\left(\frac{12K}{\alpha - \gamma}\right)} \int_0^q \frac{dq}{\{(1 - s^2 q^2)(1 - q^2)\}^{1/2}} \qquad (2.53a)$$

$$= \sqrt{\left(\frac{12K}{\alpha - \gamma}\right)} \operatorname{Sn}^{-1}(q, s), \qquad (2.53b)$$

$$\text{or} \quad u(\xi) = \beta + (\alpha - \beta)\operatorname{Cn}^2\left[\xi \sqrt{\left(\frac{\alpha - \gamma}{12K}\right)}, s\right] \qquad (2.54)$$

where Sn, Cn are the *Jacobian Elliptic functions*.

From (2.53b), the ξ-period P of $u(\xi)$ is given by

$$P = 2 \sqrt{\left(\frac{12K}{\alpha - \gamma}\right)} \int_0^1 \frac{dq}{\{(1 - s^2 q^2)(1 - q^2)\}^{1/2}}$$

$$= 4 \sqrt{\left(\frac{3K}{\alpha - \gamma}\right)} K(s^2), \qquad (2.55)$$

where $K(s^2)$ is the *complete Elliptic Integral* of the first kind. We note in this case that the bounded solution of the KdV equation represents a periodic wave whose period we could define. On account of the presence of the Cn function in the value of $u(\xi)$, the wave is called the *Cnoidal* wave.

Case (B); $\gamma = \beta \neq \alpha$.

Here u has to be confined between γ and α for the solution to be bounded and real. From eqns (2.42) and (2.45), we have, when $\gamma = \beta$,

$$d\xi = \sqrt{(3K)} \frac{du}{(u - \gamma)\sqrt{(\alpha - u)}}, \qquad (2.56)$$

which on integration yields

$$u(\xi) = \gamma + (\alpha - \gamma)\operatorname{sech}^2\left\{\sqrt{\left(\frac{\alpha - \gamma}{12K}\right)}\xi\right\}. \tag{2.57}$$

We have assumed the maximum value of u to be at the moving point where

$$\xi = x - \frac{2\gamma + \alpha}{3}t \tag{2.58}$$

vanishes and the period P of the wave is given from (2.55) by

$$P = 4\sqrt{\left(\frac{3K}{\alpha - \gamma}\right)}\int_0^1 \frac{dq}{1 - q^2} = \infty. \tag{2.59}$$

When $\xi \to \pm \infty$, from (2.57) we have

$$u \to \gamma.$$

Thus $\gamma = u_\infty$ denotes the uniform state when $\xi \to \pm \infty$. Moreover, we can denote $\alpha - \gamma$ by a and interpret a as the amplitude of the wave. We shall, therefore, preferably write eqn (2.57) as

$$u(\xi) = u_\infty + a\operatorname{sech}^2\left[\sqrt{\left(\frac{a}{12K}\right)}\left\{x - \left(u_\infty + \frac{a}{3}\right)t\right\}\right]. \tag{2.60}$$

We note the following important points about the above solution:

(i) The velocity of the wave relative to the uniform state at infinity is proportional to the amplitude a. This is the result of the nonlinearity of the wave.

(ii) The width of the wave $2\pi\sqrt{\left(\frac{12K}{a}\right)}$ is inversely proportional to the square root of amplitude.

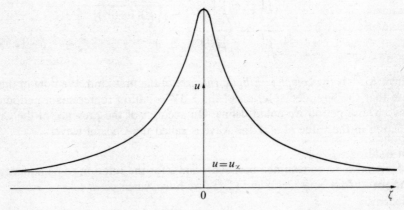

FIG. 2.7. Profile of a solitary wave joining constant states at $\xi = \pm \infty$ and localized in ξ.

(iii) The width of the wave is proportional to the square root of K; the role of K is to spread the wave. The dispersion of the nonlinear wave is exhibited through this spreading of the wave.

(iv) The amplitude is independent of the uniform state u_∞ at $\xi = \pm \infty$.

In the progressive wave represented by eqn (2.60) the transition from the constant state at $\xi \to -\infty$ to the same constant state at $\xi \to \infty$ is localized in ξ as shown by the wave profile in Fig. 2.7. We shall call such a wave a *solitary wave*.

Case C; $\gamma \neq \beta = \alpha.$

Here $f(u) \geq 0$ when $u \leq \gamma$ and we can easily show that in this case no bounded solution is possible.

We shall consider the case when $\beta \to \alpha$ but not exactly equal to α. Since now $s^2 \to 0$, we immediately deduce from Case A that

$$u(\xi) \simeq \alpha - (\alpha - \beta)\sin^2\left[\sqrt{\left(\frac{\alpha - \gamma}{12K}\right)}(\xi - \xi_0)\right] \qquad (2.61)$$

which in the limit tends to the constant solution α and period

$$P \to 2\pi\sqrt{\left(\frac{3K}{\alpha - \gamma}\right)}. \qquad (2.62)$$

Therefore, we treat the case $\beta \to \alpha$ as the limiting case of a sinusoidal wave with finite period given by eqn (2.62).

2.5. Solitary wave: solitons

In the next chapter, we shall study the most astounding property of the solitary wave, namely that two distinct solitary waves (i.e. solitary waves with distinct amplitudes and therefore with distinct velocities) interact according to the *nonlinear* KdV equation and yet emerge out of the interaction without any change of form, a property possessed by the linear waves. This behaviour of the solitary waves also resembles the behaviour of smooth and rigid particles during collisions. Consequently, the name *solitons* has been given to these solitary waves.

The solitons can be of great practical use. Let us consider a pulse carrying a bit of information with it. If the pulse suffers heavy dissipation, it may not reach the destination with appreciable intensity. Similarly, if the pulse suffers a significant dispersion, the pulse, on reaching the destination, may be so spread out and blurred, that the information may be totally unintelligible. However, if the pulse travels as a soliton, it can carry the information over long distances without being distorted and without suffering any significant loss in its intensity.

The earliest recorded evidence of existence of solitary waves is in the paper of Scott-Russell (1844). We reproduce here his interesting and exciting description of the solitary wave which he observed accidentally:

I was observing the motion of a boat which was rapidly drawn along a narrow channel by a pair of horses, when the boat suddenly stopped — not so the mass of water in the channel which it had put in motion; it accumulated round the prow of the vessel in a state of violent agitation, then suddenly leaving it behind, rolled forward with great velocity assuming the form of a large solitary elevation, a rounded, smooth and well-defined heap of water, which continued its course along the channel apparently without change of form or diminution of speed. I followed it on horseback, and overtook it still rolling at a rate of some eight to nine miles an hour, preserving its original figure some thirty feet long and a foot to a foot and a half in height. Its height gradually diminished, and after a chase of one or two miles I lost it in the windings of the channel. Such, in the month of August, 1834, was my first chance interview with that singular and beautiful phenomenon.

2.6. Some other equations of evolution exhibiting solitons

There are other equations of evolution which admit soliton solutions. We record some of them now.

2.6.1. *Generalized forms of the KdV equation*

The generalized form of the KdV equation

$$\phi_t + \alpha \phi^p \phi_x + \underbrace{\phi_{x \ldots x}}_{2r + 1 \text{ times}} = 0,$$

where $\alpha = $ constant and p, r are non-negative integers, has the following important particular cases which admit solitary wave solutions:

(i) $r = 0, p = 0$: a linear dispersionless equation which admits a solitary wave solution;

(ii) $r = 1$, $p = $ an odd integer: the equation admits a solitary wave solution with sgn (amplitude of wave) $= $ sgn (α);

(iii) $r = 1, p = $ an even integer: the equation admits *either* a compressive solitary wave solution with sgn (amplitude of wave) $= $ sgn (α) *or* a rarefaction wave solution with sgn (amplitude of wave) $= -$ sgn (α); (iv) $r = 1$, $p = 2$: the equation describes an acoustic wave in certain anharmonic lattices (Zabusky 1967) and an Alfvén wave in a collisionless plasma (Kakutani and Ono 1973); and

(v) regularized KdV equation:

$$\phi_t + \phi_x + \phi \phi_x - \phi_{xxt} = 0$$

(Benjamin, Bona, and Mahony 1972).

2.6.2. *The Boussinesq equation*

$$\phi_{xx} - \phi_{tt} + 6(\phi^2)_{xx} + \phi_{xxxx} = 0$$

describes shallow-water waves propagating in both directions (Toda and Wadati 1973, Prasad and Ravindran 1977), one-dimensional nonlinear lattice wave (Zabusky 1967). Recently, Hirota (1973) showed by numerical calculations that it possesses soliton solutions.

2.6.3. The Sine–Gordon equation

$$\phi_{xx} - \phi_{tt} = \sin\phi$$

describes the propagation of a crystal dislocation (Frenkel and Kontorova 1939, Kochendörfer and Seeger 1950, 1951, and Seeger, Donth, and Kochendörfer 1953); the Bloch wall motion of magnetic crystals (Bean and de Blois 1959, Döring 1948, and Menikoff 1972); the propagation of a 'splay wave' along a lipid membrane (Fergason and Brown 1968); a unitary theory for elementary particles (Enz 1963, Rosen and Rosenstock 1952 and Skyrme 1958, 1961); and the propagation of magnetic flux on a Josephson line (Kulik 1967, Lebwohl and Stephen 1967, Scott 1967, Scott and Johnson 1969, and Scott 1970).

2.6.4. Nonlinear lattice equation

$$m\frac{\mathrm{d}^2 y_n}{\mathrm{d}t^2} = a[\exp(br_n) - \exp(-br_{n+1})], \quad n = 1, 2; a, b = \text{constant},$$

$$r_n = y_n - y_{n-1}$$

are called the Toda lattice equations (Toda 1967a and b, 1968, 1969, 1970, Toda and Wadati 1973) and describe the motion of a one-dimensional lattice of mass points interacting through a nonlinear potential

$$\psi(r_n) = \frac{a}{b}\exp(-br_n) + ar_n - \frac{a}{b}$$

which can go from harmonic limit (ab finite, $a \to \infty, b \to 0$) to the hard-sphere limit (ab finite, $b \to \infty, a \to 0$).

2.6.5. Nonlinear Schrödinger equation

$$\phi_{xx} + i\phi_t + K|\phi|^2\phi = 0$$

describes the stationary two-dimensional self-focusing of a plane wave (Kelley 1965, Talanov 1965, and Bespalov and Talanov 1966); the one-dimensional self-modulation of a monochromatic wave (Taniuti and Washimi 1968, Asano, Taniuti, and Yajima 1969, Karpman and Kruskal 1969, and Hasegawa and Tappert 1973); the self-trapping phenomena of nonlinear optics (Karpman and Kruskal 1969); the propagation of a heat pulse in a solid (Tappert and Varma 1970); Langmuir waves in plasma (Fulton 1972, Ichikawa, Imamura, and Taniuti 1972, and Shimizu and Ichikawa 1972); and is related to the Ginzburg–Landau equation of super-conductivity (de Gennes 1966).

Derivation of the nonlinear Schrödinger equation from the Korteweg–de Vries equation and vice-versa has been discussed by Ravindran and Prasad 1979)

2.6.6   The Hirota equation

$$i\phi_t + i3\alpha|\phi|^2\phi_x + \rho\phi_{xx} + i\sigma\phi_{xxx} + \delta|\phi|^2\phi = 0,$$

where $\alpha\rho = \sigma\delta.$

This equation reduces to the nonlinear Schrödinger equation ($\alpha = \sigma = 0$) and to the modified KdV equation $\phi_t + \alpha\phi^2\phi_x + \phi_{xxx} = 0 (\rho = \delta = 0)$, and when $\alpha = \delta = 0$ it reduces to the linear equation

$$i\phi_t + \rho\phi_{xx} + i\sigma\phi_{xxx} = 0.$$

Hirota (1973) has obtained the N-soliton solution of this equation.

2.6.7.   The Born–Infeld equation

$$\phi_{xx}(1 - \phi_t^2) + 2\phi_x\phi_t\phi_{xt} - (1 + \phi_x^2)\phi_{tt} = 0.$$

Born and Infeld had obtained this equation in three spatial dimensions as a nonlinear modification of the Maxwell equations to permit the electron to appear in a natural way as a singularity (Born and Infeld 1934, 1935, Feenberg 1935, Olsen 1972, and Porter 1972).

2.6.8.   The self-induced transparency equation

McCall and Hahn (1965) discovered, through numerical computations, that ultrashort pulses of light can travel through a resonant two-level optical medium as if it were transparent. This effect has been extensively studied (McCall and Hahn 1967, 1969) and can be physically explained as follows. The time interval of an ultrashort pulse (10^{-9}–10^{-19} S) is less than the phase memory time of the atomic levels in the optical medium. Consequently, the induced polarization can retain a definite phase relationship with the incident pulse. The leading edge of the pulse then inverts the atomic population, while the trailing edge returns it to the ground state through stimulated emission. Thus the energy transferred from the leading edge of the pulse to the quantum system is recaptured by the trailing edge. Under proper conditions of coherence and intensity, the net result is a steady pulse profile which propagates without attenuation at a velocity that can be two or three orders of magnitudes less than the phase velocity of light in the medium.

Let us consider an assembly of quantum-two-level atoms. The light wave polarizes the atoms which collectively become a source of electromagnetic field. Let the atoms be distributed with uniform density n_0 and the electric field is

$$E(x,t) = E(x,t)\cos(k_0 x - \omega_0 t),$$

where the envelope $E(x,t)$ is assumed to be slowly varying with respect to the carrier $\cos(k_0 x - \omega_0 t)$. Then Maxwell's equations reduce to

$$E_t + E_x = \langle P \rangle$$

taking for convenience the velocity of light and other physical parameters to be unity. For the general case, the reader may consult the excellent survey paper by Lamb (1971).

Let us consider a single two-level atom with energy levels separated by $\omega = \omega_0 - \Delta\omega$, and denote the polarization at (x,t) due to the atom by $p(\Delta\omega,x,t)$. Now $p(\Delta\omega,x,t)$ may be approximately decomposed into components in phase and in quadrature with the electromagnetic carrier wave:

$$p(\Delta\omega,x,t) = P(\Delta\omega,x,t)\sin(k_0 x - \omega_0 t) + Q(\Delta\omega,x,t)\cos(k_0 x - \omega_0 t).$$

The Schrödinger equation for the atom reduces to the Bloch equations for the envelope functions P and Q:

$$P_t = EM - \Delta\omega Q$$
$$Q_t = \Delta\omega P$$
$$M_t = -EP,$$

where $M(\Delta\omega,xt)$ denotes the population inversion for a single atom. We can now define

$$\langle P \rangle = n_0 \int_{-\infty}^{\infty} g(\Delta\omega) P(\Delta\omega,x,t) \mathrm{d}\Delta\omega,$$

where n_0 is the atomic density and $g(\Delta\omega)$ describes the uncertainty in the energy level.

The solution of the above set of equations has been shown both experimentally and by numerical computations to break into a sequence of isolated coherent optical pulses.

Bibliography

Burgers equation

AMES, W. F. *Nonlinear partial differential equations in engineering.* Academic Press, New York (1965).

BATEMAN, H. Some recent researches on the motion of fluids. *Mon. Weath. Rev.* **43**, 163–70 (1915).

BENTON, E. R. and PLATZMAN, G. W. A table of solutions of the one-dimensional Burgers equation. *Q. Appl. Math.* **30**, 195–212 (1972).

BURGERS, J. M. Mathematical examples illustrating relations occuring in the theory of turbulent fluid motion. *Trans. R. Neth. Acad. Sci.* **17**, 1–53 (1939).
——. Application of a model system to illustrate some points of the statistical theory of free turbulence. *Proc. R. Neth. Acad. Sci.* **43**, 2–12 (1940).
CHU, C. W. A class of reducible systems of quasi-linear partial differential equations. *Q. Appl. Math.* **23**, 275–8 (1965).
COLE, J. D. On a quasi-linear parabolic equation occuring in aerodynamics. *Q. Appl. Math.* **9**, 225–36 (1951).
HOPF, E. The partial differential equation $u_t + uu_x = \mu u_{xx}$. *Communs pure appl. Math.* **3**, 201–30 (1950).
LEIBOVICH, S. and SEEBASS, R. *Nonlinear Waves,* Chapter 4. Cornell University Press, New York (1974).
LIGHTHILL, M. J. Viscosity effects in sound waves of finite amplitude, in *Surveys in mechanics* (eds. G. K. Batchelor and R. M. Davies) pp. 250–351 Cambridge University Press (1956).
SHVETS, M. E. and MELESHKO, V. P. Numerical algorithm of a solution of the system of equations of hydrodynamics of the atmosphere. *Izv. Acad. Sci., USSR. Atmospher. Ocean Phys.*, **1**, 519–20 (1965).

Korteweg–de Vries equation

BENJAMIN, T. B. The stability of solitary waves. *Proc. R. Soc.* A **328**, 153–83 (1972).
——BONA, J. L., and MAHONY, J. J. 1972: Model equations for long waves in nonlinear dispersive systems. *Phil. Trans. R. Soc.* A **272**, 47–78 (1972).
BEREZIN, Y. A. and KARPMAN, V. I. Nonlinear evolutions of disturbances in plasmas and other dispersive media. *Soviet Phys. JETP* **24**, 1049–56 (1967).
JEFFREY, A. and KAKUTANI, T. 1972: Weak nonlinear dispersive waves: a discussion centered around the Korteweg–de Vries equation. *SIAM* Rev. **14**, 582–643 (1972).
KADOMTSEV, B. B. and KARPMAN, V. I. Nonlinear waves. *Soviet Phys. Usp.* **14**, 40–60 (1971).
MENIKOFF, A. The existence of unbounded solutions of the Korteweg–de Vries equation. *Communs pure appl. Math.* **25**, 407–32 (1972).
MIURA, R. M. GARDNER, C. S., and KRUSKAL, M. D. Korteweg–de Vries equation and generalizations. II. Existence of conservation Laws and constants of the motion. *J. Math. Phys.* **9**, 1204–9 (1968).
SCOTT, A. C. CHU, F. Y. E., and MCLAUGHLIN, D. W. The soliton: a new concept in applied science, *Proc. Instn. elec. Engrs.* **61**, 1443–83 (1973).
SCOTT-RUSSELL, J. Report on waves. *Proc. R. Soc. Edinb.* 319–20 (1844).
SJOBERG, A., On the Korteweg–de Vries equation, existence and uniqueness. *J. Math. Analysis Applic.* **29**, 569–79 (1970).
TANIUTI, T. and WEI, C. C. Reductive perturbation method in nonlinear wave propagation. I. *J. Phys. Soc. Japan* **24**, pp. 941–66 (1968).
ZABUSKY, N. J. and KRUSKAL, M. D. Interaction of solitons in a collisionless plasma and the recurrence of initial states, *Phys. Rev. Lett.* **15**, 240–3 (1965).

Generalized Korteweg–de Vries Equations

BENJAMIN, T. B. BONA, J. L., and MAHONY, J. J. Model equations for long waves in nonlinear dispersive systems. *Phil. Trans. R. Soc.* A **272**, 47–78 (1972).
ZABUSKY, N. J. in *Nonlinear partial differential equations* (ed. W. Ames), pp. 223–58. Academic Press, New York (1967).

Boussinesq equation

Hɪʀᴏᴛᴀ, R., Exact N-soliton solutions of the wave equation of long waves in shallow-water and in nonlinear lattices. *J. Math. Phys.* **14**, 810–14 (1973).

Pʀᴀsᴀᴅ, P. and Rᴀᴠɪɴᴅʀᴀɴ, R. A theory of nonlinear waves in multi-dimensions: with special reference to surface water waves. *J. Inst. Math. and its Appl.* **20**, 9–20 (1977).

Tᴏᴅᴀ, M. and Wᴀᴅᴀᴛɪ, M. A soliton and two solitons in an exponential lattice and related equations. *J. phys. Soc. Japan* **34**, 18–25 (1973).

Zᴀʙᴜsᴋʏ, N. J. in *Nonlinear waves* (ed. W. Ames), pp. 223–58: Academic Press, New York (1967).

Sine – Gordon equation

Bᴇᴀɴ, C. P. and de Bʟᴏɪs, R. W. Ferromagnetic domain wall as a pseudorelativistic entity. *Bull. Am. phys. Soc.* **4**, 53 (1959).

Dᴏ̈ʀɪɴɢ, W. Über die Trägheit der Wände zwischen *weisschen Berzirken*. *Z. Naturforsch.* **31**, 373–9 (1948).

Eɴᴢ, U. Discrete mass, elementary length, and a topological invariant as a consequence of a relativistic invariant variational principle. *Phys. Rev.* **131**, 1392–4 (1963).

Fᴇʀɢᴀsᴏɴ, J. L. and Bʀᴏᴡɴ, G. H. Liquid crystals and living systems. *J. Am. Oil Chem. Soc.* **45**, 120–7 (1968).

Fʀᴇɴᴋᴇʟ, J. and Kᴏɴᴛᴏʀᴏᴠᴀ, T. On the theory of plastic deformation and twinning. *Fiz. Zh.* **1**, 137–49 (1939).

Kᴏᴄʜᴇɴᴅᴏ̈ʀꜰᴇʀ, A. and Sᴇᴇɢᴇʀ, A. Theorie der Versetzungen in eindimensionalen Atomreihen I. Periodisch angeordnete Versetzungen. *Z. Phys.* **127**, 533–50 (1950).

Kᴜʟɪᴋ, I. O. Wave propagation in a Josephson tunnel junction in the presence of vortices and the electrodynamics of weak superconductivity. *Soviet Phys. JETP* **24**, 1307–17 (1967).

Lᴇʙᴡᴏʜʟ, P. and Sᴛᴇᴘʜᴇɴ, M. J. Properties of vortex lines in superconducting barriers. *Phys. Rev.* **163**, 376–9 (1967).

Mᴇɴɪᴋᴏꜰꜰ, A. The existence of unbounded solutions of the Korteweg–de Vries equation. *Communs pure appl. Math.* **25**, 407–32 (1972).

Rᴏsᴇɴ, N. and Rᴏsᴇɴsᴛᴏᴄᴋ, H. B. The force between particles in a nonlinear field theory. *Phys. Rev.* **85**, 257–9 (1952).

Sᴄᴏᴛᴛ, A. C. Steady propagation on long Josephson junctions. *Bull. Am. Phys. Soc.* **12**, 308–9 (1967).

——Propagation of flux on a long Josephson tunnel junction. *Nuovo Cim.* **69B**, 241–61 (1970).

——and Jᴏʜɴsᴏɴ, W. J. Internal flux motion in large Josephson junctions. *Appl. Phys. Lett.* **14**, 316–18 (1969).

Sᴇᴇɢᴇʀ, A. Dᴏɴᴛʜ, H., and Kᴏ̈ᴄʜᴇɴᴅᴏʀꜰᴇʀ, A. Theorie der Versetzungen in eindimensionalen Atomreihen. II. Beliebig angeordnete und beschleunigte Versetzungen. *Z. Phys.* **134**, 173–93 (1953).

Sᴋʏʀᴍᴇ, T. H. R. A nonlinear theory of strong interactions. *Proc. R. Soc.* A **247**, 260–78 (1958).

—— Particle states of a quantized field. *Proc. R. Soc.* A **262**, 237–45 (1961).

Nonlinear lattice equation

Tᴏᴅᴀ, M. Vibration of a chain with nonlinear interaction. *J. phys. Soc. Japan* **22**, 431–6 (1967a).

——Wave propagation in anharmonic lattices. *J. phys. Soc. Japan* **23**,501–6 (1967*b*).
——. Mechanics and Statistics of nonlinear chains. *Proceedings of the International Conference on Statistical Mechanics*, Kyoto, Japan (1968). Also, *J. phys. Soc. Japan* **26**, Supplement, 235–7 (1969).
——. Waves in nonlinear lattices. *Prog. Theor. Phys.* **45**, Supplement, 174–200 (1970).
—— and WADATI, M. A soliton and two solitons in an exponential lattice and related equations, *J. phys. Soc. Japan* **34**, 18–25 (1973).

Nonlinear Schrödinger equation

ASANO, N. TANIUTI, T., and YAJIMA, N. Perturbation method for nonlinear wave modulation, II. *J. Math. Phys.* **10**, 2020–4 (1969).
BESPALOV, V. I. and TALANOV, V. I. Filamentary structure of light beams in nonlinear liquids. *JETP Lett.* **3**, 307–10 (1966).
de GENNES, P. G. *Superconductivity of metals and alloys*, Chapter 6. Benjamin, New York (1966).
FULTON, T. A. Aspects of vortices in long Josephson junctions. *Bull. Am. phys. Soc.* **17**, 46 (1972).
HASEGAWA, A. and TAPPERT, F. Transmission of stationary nonlinear optical pulses in dispersive dielectric fibers. I. Anomalous dispersion. *Appl. Phys. Lett.* **23**, 142–4 (1973).
ICHIKAWA, V. H. IMAMURA, T., and TANIUTI, T. Nonlinear wave modulation in collisionless plasma. *J. phys. Soc. Japan* **33**, 189–97 (1972).
KARPMAN, V. I. and KRUSKAL, E. M. Modulated waves in a nonlinear dispersive media. *Soviet Phys. JETP* **28**, 277–81 (1969).
KELLEY, P. L. Self-focusing of optic beams. *Phys. Rev. Lett.* **15**, 1005–8 (1965).
RAVINDRAN, R. and PRASAD, P. A mathematical analysis of nonlinear waves in a fluid filled viscoelastic tube. *Acta. Mech.* **31**, 253–80 (1979).
SHIMIZU, K. and ICHIKAWA, V. H. Automodulation of ion oscillation modes in plasma. *J. phys. Soc. Japan* **33**, 789–92 (1972).
TALANOV, V. I. Self-focusing of wave beams in nonlinear media. *JETP Lett.* **2**, 138–41 (1965).
TANIUTI, T. and WASHIMI, H. Self trapping and instability of hydromagnetic waves along the magnetic field in a cold plasma. *Phys. Rev. Lett.* **21**, 209–12 (1968).
TAPPERT, F. and VARMA, C. M. Asymptotic theory of self-trapping of heat pulses in solids. *Phys. Rev. Lett.* **25**, 1108–11 (1970).

Hirota equation

HIROTA, R. Exact envelope—soliton solutions of a nonlinear wave equation. *J. Math. Phys.* **14**, 805–9 (1973).

Born–Infeld equation

BORN, M. and INFELD, L. Foundations of a new field theory. *Proc. R. Soc.* A **144**, 425–51 (1934).
—— and——. On the quantization of the new field equations. Part 1. *Proc. R. Soc.* A **147**, 522–46 (1935). Part 2, *Proc. R. Soc.* A **150**, 141–66 (1935).
FEENBERG, F. On the Born–Infeld field theory of the electron. *Phys. Rev.* **47**, 148–57 (1935).
OLSEN, S. L. On the quantization of the Born–Infeld theory. *Lett. Nuovo Cim.* **5**, 745–7 (1972).

PORTER, J. R. Peeling and conservation laws in the Born–Infeld theory of electromagnetism. *Proc. Camb. phil. Soc.* **72**, 319–24 (1972).

Self-induced transparency equation

LAMB, G. L. Analytical descriptions of ultrashort optical pulse propagation in a resonant medium. *Rev. mod. Phys.* **43**, 99–124 (1971).

MCCALL, S. L. and HAHN, E. L. Coherent light propagation through an inhomogeneously broadened 2-level system. *Bull. Am. phys. Soc.* **10**, 1189 (1965).

—— and ——. Self-induced transparency by pulsed coherent light. *Phys. Rev. Lett.* **18**, 908–11 (1967).

—— and ——. Self-induced transparency. *Phys. Rev.* **183**, 457–85 (1969).

APPENDIX IIA

Equations governing duct flow and shallow water waves on uneven bed

In this appendix we present two physical problems and cast their governing equations in a general form. In the following appendix we indicate the method of reducing this general form to the model form. We may mention that the first problem leads to the generalized form of the Burgers equation, while the second leads to the generalized form of the KdV equation. We have considered, in Chapter 1 and 2, the propagation of waves in the homogeneous media. We have intentionally taken these problems, which deal with wave propagation in the inhomogeneous media, simply to emphasize the point that, even in the case of wave propagation in inhomogeneous media, the model equations with variable coefficients can adequately represent the physical situation.

Flow of a gas in a duct

Let us consider the wave propagation in a fluid streaming through a duct of varying cross-section. We assume that all flow variables have the same value in a cross-section of the tube so that the motion is quasi-one-dimensional. The following are the governing equations (Jeffrey and Taniuti 1964).

Continuity equation:

$$(\rho\sigma)_t + (u\rho\sigma)_x = 0 \tag{IIA.1}$$

where σ is the cross-section of the duct at the location x, ρ is the density and u the velocity of the fluid.

Momentum equation:

$$u_t + uu_x + \frac{1}{\rho}p_x - \frac{1}{\rho}\mu u_{xx} = 0, \tag{IIA.2}$$

where p is the pressure and μ is the coefficient of viscosity.

Energy equation:

$$(\omega\sigma)_t + (q\sigma)_x = 0, \tag{IIA.3}$$

where

$$\omega = \tfrac{1}{2}\rho u^2 + \frac{1}{\gamma - 1}p \tag{IIA.4}$$

and is the energy density, and

$$q = (\omega + p)u - \mu u u_x - \chi T_x \tag{IIA.5}$$

and is the energy flux, γ is the ratio of specific heats treated as constant, χ is the thermal conductivity.

Equation of state

We shall assume the following relation between pressure, density, and temperature:

$$p = p(\rho, T) \tag{IIA.6}$$

and regard T as expressible in terms of p and ρ.

We find it convenient to take $\rho\sigma = \bar{\rho}, u$ and $p\sigma = \bar{p}$ as the dependent variables and denote them as a column vector U:

$$U = \begin{bmatrix} \bar{\rho} \\ u \\ \bar{p} \end{bmatrix} \tag{IIA.7}$$

Similarly, we shall take

$$\bar{\mu} = \mu\sigma \quad \text{and} \quad \bar{\chi} = \chi\sigma. \tag{IIA.8}$$

Adopting mean values for them, we shall treat them as constant.

In terms of barred variables and barred parameters, the governing equations reduce to

$$\bar{\rho}_t + u\bar{\rho}_x + \bar{\rho}u_x = 0, \tag{IIA.9}$$

$$u_t + uu_x + \frac{1}{\bar{\rho}}\bar{p}_x - \frac{\bar{p}}{\bar{\rho}}S_x - \frac{\bar{\mu}}{\bar{\rho}}u_{xx} = 0, \tag{IIA.10}$$

and

$$\begin{aligned}
&\bar{p}_t + u\bar{p}_x + \gamma\bar{p}u_x + (\gamma - 1)u\bar{p}S_x - \bar{\mu}(\gamma - 1)u_x^2 - (\gamma - 1)\bar{\mu}_x uu_x \\
&\quad - (\gamma - 1)\bar{\chi}[\bar{\rho}_x^2 T_{\bar{\rho}\bar{\rho}} + 2\bar{\rho}_x\bar{p}_x T_{\bar{p}\bar{\rho}} + \bar{p}_x^2 T_{\bar{p}\bar{p}} + T_{\bar{\rho}}\bar{\rho}_{xx} + T_{\bar{p}}\bar{p}_{xx}] \\
&\quad - (\gamma - 1)\bar{\chi}_x[T_{\bar{\rho}}\bar{\rho}_x + T_{\bar{p}}\bar{p}_x] = 0. \tag{IIA.11}
\end{aligned}$$

A careful examination of the above equations suggests that they can be cast in the following *general* form:

$$U_t + AU_x + \sum_{\beta = 1}^{s} \prod_{\alpha = 1}^{p} \left(H_\alpha^\beta \frac{\partial}{\partial t} + K_\alpha^\beta \frac{\partial}{\partial x}\right)U + BS_x = 0, \tag{IIA.12}$$

where $H_\alpha^\beta = 0$ since in the remaining terms of the above equations there are no t-derivatives left after writing the first term; α runs over the values 1 and 2 because only first- and second-order x-derivatives are left ($p = 2, s = 3$),

$$A = \begin{bmatrix} u & \bar{\rho} & 0 \\ 0 & u & 1/\bar{\rho} \\ 0 & \gamma\bar{p} & u \end{bmatrix}, \quad S = \ln \sigma, \quad B = \begin{bmatrix} 0 \\ -\bar{p}/\bar{\rho} \\ (\gamma - 1)u\bar{p} \end{bmatrix},$$

$$K_1^1 = \begin{bmatrix} 0 & 0 & 0 \\ 0 & 0 & 0 \\ C_1 & C_2 & C_3 \end{bmatrix}, \quad K_2^1 = \begin{bmatrix} \bar{\rho} & 0 & 0 \\ 0 & u & 0 \\ 0 & 0 & \bar{\rho} \end{bmatrix},$$

$$K_1^2 = \begin{bmatrix} 0 & 0 & 0 \\ 0 & 0 & 0 \\ C_4 & 0 & C_4 \end{bmatrix}, \quad K_2^2 = \begin{bmatrix} \bar{p} & 0 & 0 \\ 0 & 0 & 0 \\ 0 & 0 & \bar{\rho} \end{bmatrix},$$

$$K_1^3 = \begin{bmatrix} 0 & 0 & 0 \\ 0 & 0 & 0 \\ C_5 - C_1\bar{\rho} - C_4\bar{p} & -C_2 u & C_6 - C_3\bar{p} - C_4\bar{\rho} \end{bmatrix}, \quad K_2^3 = I,$$

I being the unit matrix of order three and

$$C_1 = -(\gamma - 1)\bar{\chi}T_{\bar{\rho}\bar{\rho}}, \; C_2 = -(\gamma - 1)\bar{\mu}$$
$$C_3 = -(\gamma - 1)\bar{\chi}T_{\bar{p}\bar{p}}, \; C_4 = -(\gamma - 1)\bar{\chi}T_{\bar{p}\bar{\rho}},$$
$$C_5 = -(\gamma - 1)\bar{\chi}T_{\bar{\rho}}, \; C_6 = -(\gamma - 1)\bar{\chi}T_{\bar{p}}.$$

We note that we have neglected the variations in $\bar{\mu}, \bar{\chi}, T_{\bar{\rho}}, T_{\bar{p}}, T_{\bar{\rho}\bar{\rho}}, T_{\bar{\rho}\bar{p}}, T_{\bar{p}\bar{p}}$, with respect to x.

Shallow water wave propagation on an uneven bed

For long waves on a beach, Peregrine (1968) has obtained the following set of equations

$$\frac{\partial \mathbf{u}}{\partial t} + (\mathbf{u} \cdot \nabla)\mathbf{u} + \nabla h + \tfrac{1}{6}H^2\frac{\partial}{\partial t}\nabla(\nabla \cdot \mathbf{u}) - \tfrac{1}{2}H\frac{\partial}{\partial t}\nabla[\nabla \cdot (H\mathbf{u})] = 0, \quad \text{(IIA.13a)}$$

$$\frac{\partial h}{\partial t} + \nabla \cdot [(H + h)\mathbf{u}] = 0, \quad \text{(IIA.13b)}$$

where $\mathbf{u}$ is the horizontal velocity averaged over the vertical direction, H is the depth of still water which is a function of horizontal coordinates, and h is the wave amplitude.

If we consider the motion in the direction of the x-axis only, these equations reduce to

$$h_t + uh_x + (H + h)u_x + uH_x = 0 \quad \text{(IIA.14a)}$$

and

$$u_t + uu_x + h_x - \tfrac{1}{2}H(Hu_t)_{xx} + \tfrac{1}{6}H^2 u_{txx} = 0. \qquad \text{(IIA.14b)}$$

We note that (IIA.14a) contains only the first-order x- and t- derivatives but (IIA.14b) contains the third-order derivatives out of which the x-derivation is repeated twice while the t-derivation occurs only once. These remarks are helpful in properly choosing the matrices H_α^β and K_α^β. Clearly, here $p = 3$ when we cast (IIA.14) into the form (IIA.12). We choose

$$U = \begin{bmatrix} h \\ u \end{bmatrix};$$

then we choose $s = 1$,

$$A = \begin{bmatrix} u & H+h \\ 1 & u \end{bmatrix}, \qquad S = H, \qquad B = \begin{bmatrix} u \\ 0 \end{bmatrix},$$

$$H_1^1 = H_2^1 = 0, \qquad H_3^1 = \begin{bmatrix} 0 & H \\ 0 & 1 \end{bmatrix},$$

$$K_1^1 = \begin{bmatrix} 0 & 0 \\ -\tfrac{1}{2}H & \tfrac{1}{6}H^2 \end{bmatrix}, \qquad K_2^1 = I, \qquad K_3^1 = 0$$

to reduce the set of equations (IIA.14) to the form (IIA.12).

The above two examples of wave propagation in inhomogeneous media have been considered by Asano and Ono (1971).

In addition to these two examples, they have also considered the oblique magneto-acoustic wave propagation. Taniuti and Wei (1968) have considered two examples of wave propagation in homogeneous media, namely waves in moving gas and ion-acoustic waves. Using the method of singular perturbations, they have developed a neat theory of reducing a given set of equations in the standard form (IIA.12) without the last term, i.e., ignoring the inhomogeneity of media, to a single nonlinear partial differential equation under the fundamental assumptions that the nonlinearity is weak, dispersive or diffusive effects are moderate, and waves have long wavelength. We shall discuss this reduction theory as modified by Asano and Ono (1971) to take account of moderate inhomogeneity in Appendix IIB.

Other investigators have also used the method of 'multiple scales' in establishing the Burgers and the KdV equations, for example, Leibovich and Seebass (1972). Their method is, in fact, also a modification of the procedure of Taniuti and Wei (1968). Prasad and Ravindran (1977) have developed a more general method to derive similar model equations for the propagation of curved waves in multi-dimensions.

Bibliography

ASANO, N. and ONO, H. Nonlinear dispersive or dissipative waves. *J. phys. Soc. Japan* **31**, 1830–6 (1971).

JEFFREY, A. and TANIUTI, T. *Nonlinear wave propagation.* Academic Press, New York (1964).

LEIBOVICH, S. and SEEBASS, A. R. *Nonlinear waves*, Chapter IV. Cornell University Press, Ithaca (1972).

PEREGRINE, D. H. Long waves on a beach. *J. Fluid Mech.* **27**, 815–27 (1967).

PRASAD, P. and RAVINDRAN, R. A theory of nonlinear waves in multi-dimensions: with special reference to surface water waves. *J. Inst. Maths and its Appl.* **20**, 9–20 (1977).

TANIUTI, T. and WEI, C. C. Reductive perturbation method in nonlinear wave propagation-I. *J. phys. Soc. Japan*, **241**, 941–6 (1968).

APPENDIX IIB
Reduction theory

We shall first obtain the scales of stretching the coordinates by considering a comparatively simple case of a system of equations defining moderately nonlinear-wave propagation in a *homogeneous* medium, namely

$$U_t + AU_x + \sum_{\beta=1}^{s} \prod_{\alpha=1}^{p} \left(H_\alpha^\beta \frac{\partial}{\partial t} + K_\alpha^\beta \frac{\partial}{\partial x} \right) U = 0. \qquad \text{(IIB.1a)}$$

We shall later on consider a general set of equations defining moderately nonlinear-wave propagation in a moderately inhomogeneous medium:

$$U_t + AU_x + \sum_{\beta=1}^{s} \prod_{\alpha=1}^{p} \left(H_\alpha^\beta \frac{\partial}{\partial t} + K_\alpha^\beta \frac{\partial}{\partial x} \right) U + BS_x = 0, \qquad \text{(IIB.1b)}$$

in which the last term takes account of the inhomogeneity of the medium as shown in Appendix. IIA.

Here U is a column vector of n components $u_1, u_2, \ldots, u_n (n \geq 2)$, $A, H_\alpha^\beta, K_\alpha^\beta$, and B are $n \times n$ matrices whose elements depend on U and x in case of (IIB.1b) and simply on U in case of (IIB.1a), S is a known vector-valued function of x, $p \geq 2$, and suffixes x and t denote, as usual, the space and time derivations respectively.

We can obtain the dispersion relation by considering small variations around a constant state U_0 in the form

$$U = U_0 + U_1 \exp\{i(kx - \omega t)\}. \qquad \text{(IIB.2)}$$

On substituting (IIB.2) in (IIB.1a) and retaining only the first power of U_1, we have

$$\left\{ -\frac{\omega}{k}I + A_0 + i^{p-1}k^{p-1} \sum_{\beta=1}^{s} \prod_{\alpha=1}^{p} \left(K_{\alpha 0}^\beta - \frac{\omega}{k}H_{\alpha 0}^\beta \right) \right\} U_1 = 0, \qquad \text{(IIB.3)}$$

where suffix zero denotes the value when $U = U_0$.

Eqn (IIB.3) immediately leads to the dispersion relation

$$\left| -\frac{\omega}{k}I + A_0 + i^{p-1}k^{p-1} \sum_{\beta=1}^{s} \prod_{\alpha=1}^{p} \left(K_{\alpha 0}^\beta - \frac{\omega}{k}H_{\alpha 0}^\beta \right) \right| = 0. \qquad \text{(IIB.4)}$$

Since we are interested in *long* waves, we solve eqn (IIB.3) by the method of successive approximations by treating k small. The zeroth order approximation is given by

$$\left(-\frac{\omega}{k}I + A_0\right)U_{10} = 0. \tag{IIB.5}$$

The zeroth order dispersion relation is of order n in $\frac{\omega}{k}$. Let λ_0 be a nondegenerate eigenvalue of A_0, l_0 a left eigenvector and r_0 a right eigenvector of A_0 for λ_0. Then, to this approximation, we have

$$\frac{\omega}{k} = \lambda_0, U_{10} = r_0 \text{ (in fact some scalar multiple of } r_0\text{).} \tag{IIB.6}$$

The next approximation is obtained by substituting λ_0 for $\frac{\omega}{k}$ and r_0 for U_1 in the neglected terms. Thus, we have

$$\left\{-\frac{\omega}{k}I + A_0 + i^{p-1}k^{p-1}\sum_{\beta=1}^{s}\prod_{\alpha=1}^{p}(K_{\alpha 0}^{\beta} - \lambda_0 H_{\alpha 0}^{\beta})\right\}r_0 = 0 \tag{IIB.7}$$

so that on multiplying (IIB.7) by the left eigenvector l_0, we have

$$\frac{\omega}{k} = \lambda_0 + \frac{i^{p-1}k^{p-1}l_0}{l_0 r_0}\left\{\sum_{\beta=1}^{s}\prod_{\alpha=1}^{p}(K_{\alpha 0}^{\beta} - \lambda_0 H_{\alpha 0}^{\beta})\right\}r_0.$$

Proceeding in this manner, we can obtain higher order approximations to $\frac{\omega}{k}$ in the form

$$\frac{\omega}{k} = \lambda_0 + C_1 k^{p-1} + C_2 k^{2(p-1)} + \dots, \tag{IIB.8}$$

where

$$C_1 = \frac{i^{p-1}l_0\left\{\sum_{\beta=1}^{s}\prod_{\alpha=1}^{p}(K_{\alpha 0}^{\beta} - \lambda_0 H_{\alpha 0}^{\beta})\right\}r_0}{l_0 r_0}, \tag{IIB.9}$$

and so on.

We shall *assume* that $C_1 \neq 0$.

The characteristic curves of the reduced equation obtained by neglecting the third term in (IIB.1a) can also be written in the following form:

$$\frac{dx}{dt} = \lambda_0 + \varepsilon\lambda_1 + O(\varepsilon^2), \tag{IIB.10}$$

where ε is a small (but nonzero) parameter specifying the magnitude of *nonlinearity*.

Comparing eqns (IIB.8) and (IIB.10), we find that there can be coupling between nonlinear effects and the dispersive (or dissipative) effects at the order of ε, if

$$k \sim \varepsilon^a, \quad a = \frac{1}{p-1}. \tag{IIB.11}$$

Now (IIB.11) implies that ε^a times wavelength is of the order of unity. We should take account of this fact in defining the moving coordinate ξ:

$$\xi = \varepsilon^a(x - \lambda_0 t) \tag{IIB.12}$$

in which the wave can be described when nonlinearity is weak and wavelength is long.

Differentiating (IIB.12) with respect to x, we get

$$\frac{d\xi}{dx} = \varepsilon^a\left(1 - \frac{\lambda_0}{\dfrac{dx}{dt}}\right) = \varepsilon^a\left(1 - \frac{\lambda_0}{\lambda_0 + \varepsilon\lambda_1 + O(\varepsilon^2)}\right) = \varepsilon^{a+1}\left\{\frac{\lambda_1}{\lambda_0} + O(\varepsilon)\right\}$$

or

$$\frac{d\xi}{d\eta} = \lambda_1/\lambda_0 + O(\varepsilon),$$

where

$$\eta = \varepsilon^{a+1}x. \tag{IIB.13}$$

This defines another stretched variable η such that

$$\frac{d\xi}{d\eta} = O(1).$$

Differentiating (IIB.12) with respect to t, we have

$$\frac{d\xi}{dt} = \varepsilon^{a+1}\left[\lambda_1 + O(\varepsilon)\right] \tag{IIB.14}$$

This helps us in defining a variable τ:

$$\tau = \varepsilon^{a+1}t \tag{IIB.15}$$

such that

$$\frac{d\xi}{d\tau} = O(1).$$

We thus have two sets of stretched variables: (i) ξ, η and (ii) ξ, τ. It is clear that we should use the first set while dealing with an initial value problem, and the second set while dealing with a boundary value problem. The characteristics in terms of these stretched variables take the following form:

$$\frac{dx}{dt} = \lambda_0 + \varepsilon\lambda_0\frac{d\xi}{d\eta}, \text{ on using eqns (IIB.12) and (IIB.13)} \tag{IIB.16}$$

and

$$\frac{dx}{dt} = \lambda_0 + \frac{d\xi}{d\tau}, \text{ on using eqns (IIB.12) and (IIB.15).} \qquad \text{(IIB.17)}$$

From the foregoing discussion it is clear that the scale of coordinate stretching is uniquely determined once the set of governing equations (IIB.1a) is specified.

When the medium is not homogeneous, as is the case with most of the physical systems, the interaction of the wave with the inhomogeneity also becomes important.

We shall now apply the reduction theory to eqn (IIB.1b) based on the method of stretching the coordinates and developed by Taniuti and Wei (1968) for the case of homogeneous media and modified by Asano and Ono to take account of moderate inhomegeneity (1971).

The steady state of the system, denoted by the zero suffix, is given by

$$A_0 U_{0x} + \sum_{\beta=1}^{s} \prod_{\alpha=1}^{p} \left(K_{\alpha 0}^{\beta} \frac{\partial}{\partial x} \right) U_0 + B_0 S_x = 0. \qquad \text{(IIB.18)}$$

We assume that U_0 and S vary slowly with x and this slow variation can be adequately taken care of by the variable η defined in (IIB.13) with $a = 1/(p-1)$. By studying the variations of U_0 and S with x, we can determine the order of ε. We shall use the stretched variables ξ and η redefined by the following relations to take account of the inhomogeneity of the medium:

$$\xi = \varepsilon^a \left(\int \frac{dx}{\lambda_0} - t \right)$$

and

$$\eta = \varepsilon^{a+1} x, \qquad \qquad \text{(IIB.19)}$$

where λ_0 is a nondegenerate eigenvalue of A_0. λ_0 is, therefore, equal to the velocity of the linear wave. In terms of the stretched coordinates, eqn (IIB.18) reduces to

$$A_0 U_{0\eta} + B_0 S_\eta = 0 \qquad \text{(IIB.20)}$$

if we neglect the term of the order of ε^p.

Let us expand U about U_0 in powers of ε:

$$U = U_0 + \varepsilon U_1 + \varepsilon^2 U_2 + \dots \qquad \text{(IIB.21a)}$$

and then expand the coefficient matrices also in powers of ε as indicated below:

$$A = A_0 + \varepsilon A_1 + \dots, \qquad \text{(IIB.21b)}$$
$$B = B_0 + \varepsilon B_1 + \dots, \qquad \text{(IIB.21c)}$$
$$H_\alpha^\beta = H_{\alpha 0}^\beta + \varepsilon H_{\alpha 1}^\beta + \dots, \qquad \text{(IIB.21d)}$$

and
$$K_\alpha^\beta = K_{\alpha 0}^\beta + \varepsilon K_{\alpha 1}^\beta + \dots. \tag{IIB.21e}$$

From (IIB.19) we have

$$\frac{\partial}{\partial t} = -\varepsilon^a \frac{\partial}{\partial \xi}, \quad \frac{\partial}{\partial x} = \varepsilon^a \left(\frac{1}{\lambda_0} \frac{\partial}{\partial \xi} + \varepsilon \frac{\partial}{\partial \eta} \right) \tag{IIB.22}$$

so that we are replacing the derivation with respect to t by derivation with respect to ξ and therefore $\dfrac{\partial U_0}{\partial \xi} = 0, \dfrac{\partial S}{\partial \xi} = 0$. Transforming (IIB.1b) to ξ, η coordinates, substituting (IIB.21) in the transformed equation and separately equating the terms of orders ε, ε^2, ... to zero, we get

$$(A_0 - \lambda I)U_{1\xi} = 0, \tag{IIB.23}$$

$$-U_{2\xi} + \frac{1}{\lambda_0} A_0 U_{2\xi} + \frac{1}{\lambda_0} A_1 U_{1\xi} + A_0 U_{1\eta} + A_1 U_{0\eta}$$

$$+ \sum_{\beta=1}^{s} \prod_{\alpha=1}^{p} \left(-H_{\alpha 0}^\beta + \frac{1}{\lambda} K_{\alpha 0}^\beta \right) U_{1\underbrace{\xi\xi\dots\xi}_{p \text{ times}}} + B_1 S_\eta = 0, \tag{IIB.24}$$

and so on.

Let r_0 be a right eigenvector of A_0 for the eigenvalue λ_0, then we have
$$(A_0 - I\lambda_0)r_0 = 0. \tag{IIB.25}$$

From eqns (IIB.23) and (IIB.25), we can write
$$U_{1\xi} = r_0 \phi_\xi, \tag{IIB.26}$$

where ϕ is a scalar function of ξ, η, so that
$$U_1 = r_0 \phi(\xi, \eta) + V(\eta), \tag{IIB.27}$$

where V is an arbitrary function of η alone.

Since the arbitrariness of the norm of r_0 does not affect the final result, V is uniquely determined if U_1 and ϕ are given at some value of ξ, say $\xi = \xi_0$. The equation (IIB.24) then determines ϕ. Multiplying (IIB.24) by a left eigenvector l_0 of A_0 for the eigenvalue λ_0, we have, on some simplification and using eqns (IIB.25), (IIB.26), and (IIB.27),

$$\frac{1}{\lambda_0} l_0 A_1 r_0 \phi_\xi + \lambda_0 l_0 r_0 \phi_\eta + \lambda_0 l_0 V_\eta + l_0 A_1 U_{0\eta} + \lambda_0 l_0 r_{0\eta} \phi$$

$$+ l_0 \left\{ \sum_{\beta=1}^{s} \prod_{\alpha=1}^{p} \left(-H_{\alpha 0}^\beta + \frac{1}{\lambda_0} K_{\alpha 0}^\beta \right) \right\} r_0 \phi_{\underbrace{\xi\xi\dots\xi}_{p \text{ times}}} + l_0 B_1 S_\eta = 0. \tag{IIB.28}$$

This can be further simplified by nothing that

$$A(u_1, u_2, \ldots, u_n) = A(u_{10} + \varepsilon u_{11} + \ldots + u_{20}$$
$$+ \varepsilon u_{21} + \ldots + u_{n0} + \varepsilon u_{n1} + \ldots)$$

$$= A(U_0) + \sum_{i=1}^{n} \left(\frac{\partial A}{\partial u_i}\right)_0 \cdot \varepsilon u_{i1}$$

$$\equiv A_0 + \varepsilon A_1$$

so that

$$A_1 = (\nabla_U A)_0 \cdot U_1 = (\nabla_U A)_0 \cdot (r_0 \phi + V)$$
$$= \{(\nabla_U A)_0 \cdot r_0\} \phi + (\nabla_U A)_0 \cdot V, \tag{IIB.29}$$

with a similar expression for B_1. Substituting eqn (IIB.29) in eqn (IIB.28) and rearranging the terms, we have the equation for ϕ:

$$\phi_\eta + (\alpha\phi + \alpha')\phi_\xi + \beta\underbrace{\phi_{\xi\xi\ldots\xi}}_{p\ \text{times}} + \gamma\phi + \gamma' = 0, \tag{IIB.30}$$

where

$$\alpha = \frac{l_0\{(\nabla_U A)_0 \cdot r_0\}r_0}{\lambda_0^2 l_0 r_0}, \tag{IIB.31a}$$

$$\alpha' = \frac{l_0\{(\nabla_U A)_0 \cdot V\}r_0}{\lambda_0^2 l_0 r_0}, \tag{IIB.31b}$$

$$\beta = \frac{l_0\left\{\sum_{\beta=1}^{s} \prod_{\alpha=1}^{p} \left(-H_{\alpha 0}^\beta + \frac{1}{\lambda_0}K_{\alpha 0}^\beta\right)\right\}r_0}{\lambda_0 l_0 r_0}, \tag{IIB.31c}$$

$$\gamma = \frac{\lambda_0 l_0 r_{0\eta} + l_0[\{(\nabla_U A)_0 r_0\}U_{0\eta} + \{(\nabla_U B)_0 r_0\}S_\eta]}{\lambda_0 l_0 r_0}, \tag{IIB.31d}$$

and

$$\gamma' = \frac{\lambda_0 l_0 V_\eta + l_0[\{(\nabla_U A)_0 V\}U_{0\eta} + \{(\nabla_U B)_0 V\}S_\eta]}{\lambda_0 l_0 r_0}. \tag{IIB.31e}$$

We have thus reduced the consideration of the given set of equations (IIB.1b) to the consideration of single equation in ϕ. In passing we note that $\beta \neq 0$ under our assumption that $C_1 \neq 0$ (eqn. (IIB.9)). Taniuti and Wei (1968), and Asano and Ono (1971) have discussed the case $C_1 = 0$ in their papers.

We can remove a number of terms from (IIB.30) with the help of the following transformation

$$\psi = \exp\left(\int \gamma \mathrm{d}\eta\right)\phi + \int \gamma' \exp\left(\int \gamma \mathrm{d}\eta\right)\mathrm{d}\eta, \tag{IIB.32a}$$

$$\zeta = \xi - \int \alpha' d\eta + \int \left\{ \alpha \exp\left(- \int \gamma d\eta \right) \int \gamma' \exp\left(\int \gamma d\eta \right) d\eta \right\} d\eta, \quad \text{(IIB.32b)}$$

and

$$\theta = \eta. \quad \text{(IIB32c)}$$

The above transformation takes

$$(\xi, \eta) \to (\zeta, \theta) \quad \text{(IIB.33a)}$$

and

$$\phi \to \exp\left(- \int \gamma d\theta \right) \left\{ \psi(\zeta, \theta) - \int \gamma' \exp\left(\int \gamma d\theta \right) d\theta \right\} \quad \text{(IIB.33b)}$$

and the derivatives transform according to the following scheme:

$$\frac{\partial}{\partial \xi} = \frac{\partial}{\partial \zeta} \quad \text{(IIB.33c)}$$

and

$$\frac{\partial}{\partial \eta} = \frac{\partial}{\partial \theta} + \left[-\alpha' + \alpha \exp\left(- \int \gamma d\theta \right) \int \gamma' \exp\left(\int \gamma d\theta \right) d\theta \right] \frac{\partial}{\partial \zeta}. \quad \text{(IIB.33d)}$$

The transformed equation takes the convenient form

$$\psi_\theta + \alpha \exp\left(- \int \gamma d\theta \right) \psi \psi_\zeta + \beta \underbrace{\psi_{\zeta\zeta\ldots\zeta}}_{p \text{ times}} = 0, \quad \text{(IIB.34)}$$

where the coefficients are functions of θ only.

It is clear that when $p = 2$, it reduces to the Burgers equation and when $p = 3$, it reduces to the KdV equation both with variable coefficients.

We shall now apply the above theory to the two cases discussed in Appendix IIA.

Duct flow

The steady solutions are given by

$$u_0 \bar{\rho}_{0\eta} + \bar{\rho}_0 u_{0\eta} = 0, \quad \text{(IIB.35a)}$$

$$u_0 u_{0\eta} + \frac{1}{\bar{\rho}_0} \bar{p}_{0\eta} - \frac{\bar{p}_0 \sigma_\eta}{\bar{\rho}_0 \sigma} = 0, \quad \text{(IIB.35b)}$$

and

$$\gamma \bar{p}_0 u_{0\eta} + u_0 \bar{p}_{0\eta} + (\gamma - 1) u_0 \bar{p}_0 \frac{\sigma_\eta}{\sigma} = 0. \quad \text{(IIB.35c)}$$

The eigenvalues of

$$A_0 = \begin{bmatrix} u_0 - \lambda_0 & \bar{\rho}_0 & 0 \\ 0 & u_0 - \lambda_0 & \dfrac{1}{\bar{\rho}_0} \\ 0 & \gamma\bar{p}_0 & u_0 - \lambda_0 \end{bmatrix} \qquad \text{(IIB.36)}$$

and are given by

$$\lambda_0 = u_0, \lambda_0 = u_0 \pm a_0, \text{ where } a_0^2 = \frac{\gamma p_0}{\rho_0}. \qquad \text{(IIB.37)}$$

We shall consider

$$\lambda_0 = u_0 + a_0. \qquad \text{(IIB.38)}$$

A right eigenvector r_0 and a left eigenvector l_0 of A_0 for λ_0 are

$$r_0 = \begin{bmatrix} \bar{\rho}_0 \\ a_0 \\ \gamma\bar{p}_0 \end{bmatrix}, \quad l_0 = [0, \gamma\bar{p}_0, a_0]; \qquad \text{(IIB.39)}$$

then

$$(\nabla_U A)_{U_0} \cdot r_0 = \begin{bmatrix} a_0 & \bar{\rho}_0 & 0 \\ 0 & a_0 & -\dfrac{1}{\bar{\rho}_0} \\ 0 & \gamma^2 \bar{p}_0 & a_0 \end{bmatrix} \qquad \text{(IIB.40a)}$$

and

$$l_0 r_0 = 2\gamma\bar{p}_0 a_0. \qquad \text{(IIB.40b)}$$

Therefore, we have, from (IIB.31a),

$$\alpha = \frac{\gamma + 1}{2} \frac{a_0}{\lambda_0^2}. \qquad \text{(IIB.41)}$$

Similarly, we have from (IIB.31c),

$$\beta = -\frac{1}{2\gamma\bar{p}_0\lambda_0^2} [(\gamma - 1)\bar{\chi}\{(T_{\bar{\rho}})_0\bar{\rho}_0 + (T_{\bar{p}})_0\bar{p}_0\} + \bar{\mu}a_0^2] \qquad \text{(IIB.42)}$$

and, from (IIB.31d),

$$\gamma^* = \frac{1}{2}\frac{a_{0\eta}}{a_0} + \frac{1}{2}\frac{\bar{p}_{0\eta}}{\bar{p}_0} + \frac{M_{0\eta}}{M_0 + 1}, \qquad \text{(IIB.43)}$$

where γ^* has been taken in place of γ to avoid confusion, and $M_0 = u_0/a_0$ is the Mach number of the unperturbed flow determined from (IIB.35a, b, and c). Finally, we have the Burgers equation for

$$\psi = (a_0\bar{p}_0\sigma)^{1/2}(M_0 + 1)\phi \qquad \text{(IIB.44)}$$

in the form

$$\psi_\eta + \alpha^* \psi \psi_\xi + \beta \psi_{\xi\xi} = 0, \qquad \text{(IIB.45)}$$

where

$$\alpha^* = \frac{\gamma+1}{2} \frac{a_0}{\lambda_0^3} \left(\frac{a_0}{\sigma \bar{p}_0}\right)^{1/2} \qquad \text{(IIB.46)}$$

and β are the functions of η.

Shallow water wave on uneven bed

For the constant state $h = u = 0$, the matrix

$$A_0 = \begin{bmatrix} 0 & H \\ 1 & 0 \end{bmatrix} \qquad \text{(IIB.47)}$$

so that

$$\lambda_0 = \pm H^{1/2}. \qquad \text{(IIB.48)}$$

A right eigenvector r_0 and a left eigenvector l_0 of A_0 for λ_0 are

$$r_0 = \begin{bmatrix} H \\ \lambda_0 \end{bmatrix}, \quad l_0 = [1, \lambda_0] \qquad \text{(IIB.49)}$$

and the coefficients (IIB.31) for $\lambda_0 = H^{1/2}$ are given by

$$\alpha = \frac{3}{2} \frac{1}{H^{1/2}}, \beta = \frac{1}{6} H^{1/2}, \gamma = \frac{5}{4} \frac{1}{H} \frac{\mathrm{d}H}{\mathrm{d}\eta}. \qquad \text{(IIB.50a, b and c)}$$

The KdV equation for

$$\psi = H^{5/4} \phi \qquad \text{(IIB.51)}$$

takes the form

$$\psi_\eta + (\tfrac{3}{2} H^{-7/4}) \psi \psi_\xi + (\tfrac{1}{6} H^{1/2}) \psi_{\xi\xi\xi} = 0, \qquad \text{(IIB.52)}$$

where the coefficients are now functions of η.

In passing we may mention that similar reduction theory has been used in other contexts also, for example in studying the behaviour of the partial differential equations in the neighbourhood of a critical point, [Kulikovskii and Slobodkina (1967), Bhatnagar and Prasad (1971), Prasad (1973)], and for calculating the complete history of a pulse with curved wave front [Prasad (1975)].

Bibliography

BHATNAGAR, P. L. and PRASAD, P. Study of the self-similar and steady flows near singularities, Part I. *Proc. R. Soc.* A **315**, 569–84 (1970); Part II, A **322**, 45–62 (1971).

KULIKOVSKII, A. G. and SLOBODKINA, F. A. Equilibrium of arbitrary steady flows at the transonic points, *P.M.M.*, **31**, 623–30 (1967).

PRASAD, P. Nonlinear wave propagation on an arbitrary steady transonic flow. *J. Fluid Mech.* **57**, 721–37 (1973).

——Approximation of perturbation equations of a quasi-linear hyperbolic system in a neighbourhood of a bicharacteristic. *J. Math. Analysis and its Appl.* **50**, 470–82 (1975).

3

SOLITON INTERACTION

3.1. Introduction

In Chapter 2, we have obtained the solitary wave solution

$$u(x,t) = u_\infty + a \operatorname{sech}^2\left[\sqrt{\left(\frac{a}{12K}\right)}\left\{x - \left(u_\infty + \frac{a}{3}\right)t\right\}\right] \tag{3.1}$$

for the KdV equation

$$u_t + uu_x + Ku_{xxx} = 0 \tag{3.2}$$

when two smaller roots β, γ of the cubic equation $\dfrac{du}{d\xi} \equiv f(u) = 0$ are equal.

In the present chapter, for sake of simplicity, we shall take the KdV equation in the form

$$u_t - 6uu_x + u_{xxx} = 0 \tag{3.3}$$

obtained from eqn (3.2) by the following transformation

$$x \to K^{1/3}x, \quad u \to -6K^{1/3}u, \quad t \to t. \tag{3.4}$$

We shall assume the initial condition $u(x,0) = u_0(x)$ to be bounded and thrice continuously differentiable and as in Chapter 2 take the following boundary condition:

$u(x,t)$, along with its derivatives, tends to zero as $|x| \to \infty$. $\qquad$ (3.5)

Under these boundary conditions the solitary wave solution of (3.3) is

$$u(x,t) = -\frac{a^2}{2}\operatorname{sech}^2\left\{\frac{a}{2}(x - a^2t)\right\}, \tag{3.6}$$

where a in (3.6) can be easily expressed in terms of a and K in (3.1).

The objective of the present chapter is to study the interaction between two and more solitary waves which leads to the definition of solitons. As mentioned earlier, solitons can be of immense practical utility.

3.2. Properties of the Schrödinger equation

The method, which we shall follow in studying the interaction between solitons, consists in associating the KdV equation with the steady one-dimensional Schrödinger equation. The chief merit of the method lies in the fact that we can use the properties of this celebrated equation which has been thoroughly investigated during the last thirty to forty years.

We shall record below only those properties of the Schrödinger equation which we need in our discussion. We refer the reader to Landau and Lifschitz (1958) for further reference.

We shall consider the time-independent one-dimensional Schrödinger equation

$$\psi_{xx} + \{\lambda - u(x)\}\psi = 0, \tag{3.7}$$

where $u(x)$ is the potential, λ is the energy eigenvalue and ψ is the wave function. In quantum mechanics, the function ψ in (3.7) is the wave function of a moving particle under an external field whose potential energy is $u(x)$. The following properties will be needed.

(i) The eigenvalues of λ may be discrete or continuous or both in a given problem.

(ii) The discrete eigenvalues are negative and correspond to the stable states of finite motion of the particle, where by finite motion we mean the motion which is confined to finite bounded space. We shall denote the discrete eigenvalues by

$$\lambda = -\kappa_1^2, -\kappa_2^2, \ldots, -\kappa_m^2, \text{ where } \kappa_i > 0 \tag{3.8}$$

and generally call κ_i eigenvalues.

(iii) The continuous eigenvalues correspond to infinite motion in which the particle reaches infinity. At sufficiently large distances the potential field $u(x)$ may be neglected and the particle may be regarded as free. The energy of a free particle is positive. Therefore the continuous eigenvalues are positive. We shall denote the continuous eigenvalues by

$$\lambda = k^2, \quad k > 0. \tag{3.9}$$

(iv) In classical mechanics, a particle with energy E will not be able to penetrate the region where $E < u(x)$. However, in quantum mechanics, a particle in a finite motion may be found in the regions of space where $E < u(x)$, though the probability of finding the particle in these regions is small (but nonzero) and it rapidly tends to zero as the distance into such a region increases.

(v) None of the discrete eigenvalues is degenerate, i.e., to each discrete eigenvalue, one and only one eigenfunction corresponds as seen below.

If possible, let two eigenfunctions ψ_1 and ψ_2 correspond to a discrete eigenvalue λ. Then we have

$$\frac{\psi_{1xx}}{\psi_1} = u - \lambda = \frac{\psi_{2xx}}{\psi_2}$$

or
$$\psi_{1xx}\psi_2 - \psi_{2xx}\psi_1 = 0$$

or, on integrating once,

$$\psi_{1x}\psi_2 - \psi_{2x}\psi_1 = \text{constant} = 0$$

in view of the boundary condition

$$\psi_{1,2} \to 0 \quad \text{as} \quad |x| \to \infty.$$

Therefore, we have

$$\frac{\psi_{1x}}{\psi_1} = \frac{\psi_{2x}}{\psi_2},$$

which on integration yields

$$\psi_1 = \text{constant times } \psi_2$$

i.e., the two eigenfunctions are the same except for a constant factor.

The continuous eigenvalues are degenerate.

(vi) Let us arrange the discrete eigenvalues in order of their magnitudes: $\lambda_1 < \lambda_2 < \lambda_3 < \ldots < \lambda_m$ and let ψ_n correspond to λ_n. Then the $(n+1)$th eigenfunction vanishes n times in the finite domain of the x-axis in which the motion of the particle is confined.

(vii) Let us consider a potential $u(x) \to 0$ as $|x| \to \infty$. Then, when $|x| \to \infty$, the Schrödinger equation assumes the following asymptotic form:

$$\psi_{xx} + \lambda\psi = 0.$$

For discrete eigenvalues, it takes the form

$$\psi_{xx} - \kappa^2\psi = 0$$

and its two independent solutions are

$$\psi = c_{\pm} \exp(\pm\kappa x). \tag{3.10}$$

Clearly then, when $x \to \infty$, the admissible solution is $\exp(-\kappa x)$, and when $x \to -\infty$, the admissible solution is $\exp(\kappa x)$.

For the continuous eigenvalues, the Schrödinger equation takes the form

$$\psi_{xx} + k^2\psi = 0$$

and its two independent solutions are

$$\psi = a_{\pm}(k)\exp(\pm ikx). \tag{3.11}$$

Here the solution $\exp(ikx)$ corresponds to the particle moving in the positive direction of the x-axis, while the solution $\exp(-ikx)$ corresponds to the particle moving in the negative direction of the x-axis.

(viii) In our discussion, using $u(x,t)$ determined by the KdV equation (3.3) as the potential in the Schrödinger equation, we shall treat t as a parameter with the result that the eigenvalues as well as the eigenfunctions of the Schrödinger equation will depend parametrically on t. Thus we shall write

$$\lambda = \lambda(t), \ \psi(x) = \psi(x; t). \tag{3.12}$$

(ix) When $x \to \infty$ and $u \to 0$, we can take the wave function of a continuous eigenvalue asympotically as a linear combination of two plane waves $\exp(\pm ikx)$ and similarly when $x \to -\infty$.

Let us consider a plane wave coming from $x = \infty$. We can write

$$\psi \sim \exp(-ikx) + b(k) \exp(ikx) \quad \text{as } x \to \infty$$

and
$$\psi \sim a(k) \exp(-ikx), \text{ as } x \to -\infty, \tag{3.13}$$

where the complex numbers $b(k)$ and $a(k)$ depending on the wave number are, respectively, the reflection and the transmission coefficients. Here we have taken the incident wave with amplitude unity. In case of discrete eigenvalues κ_m the eigenfunction $\psi_m \to 0$ as $|x| \to \infty$ and is square integrable, so we effect the normalization according to the following rule:

$$\int_{-\infty}^{\infty} \psi_m^2 dx = 1. \tag{3.14}$$

From the law of conservation of energy, we have:
Energy of the incident wave = Energy of the reflected wave
$$\qquad\qquad\qquad\qquad + \text{ Energy of the transmitted wave.}$$

i.e.,
$$1 = |b|^2 + |a|^2. \tag{3.15}$$

$\kappa_m, c_m, b(k)$, and $a(k)$ constitute the scattering parameters of the wave.

(x) The Schrödinger equation is of order two and therefore we shall need two independent solutions in our analysis. Let us assume that ψ is one of the solutions. Let ϕ be another linearly independent solution of the equation, then

$$\phi_{xx} + (\lambda - u)\phi = 0.$$

On substituting $\phi = \psi X$ and using the fact that

$$\psi_{xx} + (\lambda - u)\psi = 0,$$

we have

$$\frac{X_{xx}}{X_x} + \frac{2\psi_x}{\psi} = 0$$

which on integration yields

$$X_x = \frac{A}{\psi^2}.$$

Hence we have

$$X = A \int \frac{dx}{\psi^2} + B$$

and the desired other solution is

$$\phi = A\psi \int \frac{dx}{\psi^2} + B\psi$$

We can neglect the second term on the right as it will be included in the ψ-term. Therefore we shall take

$$\phi = \psi \int_0^x \frac{dx}{\psi^2}.$$

(xi) Let the potential $u(x)$ be even. Then the Schrödinger equation is invariant under the transformation

$$x \to -x.$$

Therefore, if $\psi(x)$ is an eigenfunction, $\psi(-x)$ is also an eigenfunction and, apart from a constant factor, they are same. Thus we set

$$\psi(-x) = c\psi(x).$$

Once more effecting the transformation $x \to -x$, we have

$$\psi(x) = c\psi(-x) = c^2\psi(x)$$

so that $c = \pm 1$ and $\psi(-x) = \pm \psi(x)$. Therefore, if the potential is symmetrical about $x = 0$, the wave functions of the steady states are either even or odd. This result is, in fact, the outcome of our assumption that the eigenvalues of the Schrödinger operator corresponding to the discrete eigenvalues are nondegenerate. If we do not make this assumption, under the transformation $x \to -x$, the two eigenfunctions corresponding to the same eigenvalue may simply transform into each other.

3.3. Integrals of equation and relationship between KdV equation and the Schrödinger equation

We can write the KdV equation

$$u_t - 6uu_x + u_{xxx} = 0 \tag{3.16}$$

in the conservation form

$$T_t + X_x = 0 \tag{3.17}$$

where $T = u, X = -3u^2 + u_{xx}$. If we assume that u is periodic in x or that u and its derivatives vanish sufficiently rapidly at the two infinities $x = \pm \infty$, integrating the *conservation law* (3.17) we get

$$\frac{\partial}{\partial t}\int T dx = 0 \text{ or } I \equiv \int T dx = \text{independent of time.} \tag{3.18}$$

Thus $I = \int u\,dx$, where the limits of integration are $\pm \infty$ or two ends of a period in x, is a time invariant functional of the solutions of the KdV equation. We call a time invariant functional an *integral* of equation. Miura, Gardner, and Kruskal (1968) proved that the KdV equation has an infinity of polynominal *conservation laws*. The first in this sequence is given above. The second one can be easily derived by multiplying (3.16) by u so that

$$T = \tfrac{1}{2}u^2, \quad X = \tfrac{1}{3}u^3 + uu_{xx} - \tfrac{1}{2}u_x^2. \tag{3.19}$$

The third was derived by Whitham in 1967

$$T = \tfrac{1}{3}u^3 - u_x^2, \quad X = \tfrac{1}{4}u^4 + u^2 u_{xx} - 2uu_x^2 - 2u_x u_{xxx} + u_{xx}^2. \tag{3.20}$$

Since each conservation law gives an integral of the equation, it follows that the KdV equation has an infinity of integrals of the form (3.18). However, these are not the only integrals as we shall see soon.

We shall now explain the situation in which the Schrödinger equation first appeared in the study of the KdV equation.

Let us set

$$u = v^2 + v_x \tag{3.21}$$

in the KdV equation (3.16). We thus have

$$2v(v_t - 6v^2 v_x + v_{xxx}) + (v_t - 6v^2 v_x + v_{xxx})_x = 0$$

which is satisfied when v satisfies the *associated KdV equation*

$$v_t - 6v^2 v_x + v_{xxx} = 0. \tag{3.22}$$

Thus, if v evolves according to (3.22), u defined by (3.16) evolves according to the KdV equation. The associated KdV equation can also give an infinity of polynomial conservation laws of the form (3.17), the first three being

$$T = v, \quad X = \tfrac{1}{3}v^3 + v_{xx} \tag{3.23}$$

$$T = \tfrac{1}{2}v^2, \quad X = \tfrac{1}{4}v^4 + vv_{xx} - \tfrac{1}{2}v_x^2 \tag{3.24}$$

and

$$T = \tfrac{1}{4}v^4 - \tfrac{3}{2}v_x^2, \quad X = \tfrac{1}{6}v^6 + v^3 v_{xx} - 3v^2 v_x^2 - 3v_x v_{xxx} + \tfrac{3}{2}v_{xx}^2. \tag{3.25}$$

These conservation laws give the integrals of the associated KdV equation. The equation (3.21) provides a relation between the above mentioned integrals of the KdV equation (3.16) and those of the associated KdV equation (3.22).

The relation is one to one except the fact that the integral $\int v\,dx$ of the associated KdV equation cannot be obtained from an integral of the KdV equation by the transformation (3.21).

For a given function u, the relation (3.21) becomes a Riccati equation which can be linearized by the well known transformation

$$v = \frac{\psi_x}{\psi}. \tag{3.26}$$

The equation for ψ is the one-dimensional Schrödinger equation

$$\psi_{xx} - u\psi = 0 \tag{3.27}$$

with the energy-level terms missing. Noting that the KdV equation is invariant under the transformation

$$t \to t', \; x \to x' - 6ct', \; u \to u' + c, \tag{3.28}$$

the energy levels in (3.27) can be introduced. This also leads to a proof that the eigenvalues of the Schrödinger equation

$$\psi_{xx} + (\lambda - u)\psi = 0, \tag{3.29}$$

where u evolves according to the KdV equation, are time independent functionals of u, i.e. they are integrals of the KdV equation (3.16). We shall give a straightforward proof of this in the next section.

3.4. Time-independence of the eigenvalues of the Schrödinger equation, determination of scattering parameters

We shall now prove that the discrete eigenvalues of the Schrödinger equation (3.29) are independent of t which occurs parametrically in the potential $u(x;t)$ which we have taken to be evolving according to the KdV equation which is time-dependent.

From (3.29), we have

$$u = \frac{\psi_{xx}}{\psi} + \lambda. \tag{3.30}$$

Substituting (3.30) in (3.17) we have

$$\lambda_t \psi^2 + (\psi R_x - \psi_x R)_x = 0, \tag{3.31}$$

where

$$R = \psi_t + \psi_{xxx} - 6\lambda\psi_x - 3\frac{\psi_x\psi_{xx}}{\psi} \tag{3.32a}$$

$$= \psi_t + \psi_{xxx} - 3(u + \lambda)\psi_x. \tag{3.32b}$$

Integrating (3.31) with respect to x from $x = -\infty$ to ∞ and considering the normalized wave functions, we have

$$\lambda_t = 0, \tag{3.33}$$

in view of the boundary conditions on ψ, namely ψ and its derivatives tend to zero as $|x| \to \infty$. This completes the proof of the above statement.

The most important consequence of this result is that we can once for all determine the eigenvalues of the Schrödinger equation using the initial condition $u(x,0) = u_0(x)$ *a priori* prescribed for solving the KdV equation.

For the continuous eigenvalues, λ can be assumed constant in t, hence (3.33) can be taken to be valid.

Using (3.33) in (3.31), we have for discrete as well as continuous eigenvalues:

$$\psi R_{xx} - R\psi_{xx} = 0 \tag{3.34}$$

which in view of (3.29) reduces to

$$R_{xx} + (\lambda - u)R = 0. \tag{3.35}$$

Thus R also satisfies the Schrödinger equation and we can take

$$R = C(t)\psi + D(t)\psi \int_0^x \frac{dx}{\psi^2}. \tag{3.36}$$

Let us now consider a discrete eigenvalue λ. Since $\psi \to 0$ as $|x| \to \infty$, for boundedness of R, we must take

$$D = 0 \tag{3.37}$$

and then $$R = C(t)\psi. \tag{3.38}$$

Or, from (3.32a),

$$C(t)\psi^2 = \psi\psi_t + \psi\psi_{xxx} - 6\lambda\psi\psi_x - 3\psi_x\psi_{xx}. \tag{3.39}$$

Integrating the above relation with respect to x from $x = -\infty$ to $x = \infty$ and using the boundary conditions on ψ and its derivatives and the normality condition (which says $\dfrac{\partial}{\partial t} \displaystyle\int_{-\infty}^{\infty} \psi^2 dx = 0$), we have

$$C(t) = 0 \tag{3.40}$$

so that

$$\psi_t + \psi_{xxx} - 6\lambda\psi_x - 3\frac{\psi_x\psi_{xx}}{\psi} = 0. \tag{3.41}$$

When $x \to +\infty$, we can write

$$\psi \sim c_m(t)\exp(-\kappa_m x), \quad \lambda_m = -\kappa_m^2, \quad \kappa_m > 0. \tag{3.42}$$

Substituting (3.42) in (3.41), we have

$$\frac{d}{dt}c_m = 4\kappa_m^3 c_m$$

which on integration yields

$$c_m(\kappa_m, t) = c_m(\kappa_m, 0) \exp(4\kappa_m^3 t). \tag{3.43}$$

We shall now consider the continuous eigenvalue k. For a steady plane wave coming from $x = \infty$, we can write

$$\psi \sim a(k,t) \exp(-ikx), \; x \to -\infty \tag{3.44}$$

where a is the transmission coefficient.

Substituting (3.44) in (3.36) where R is given by (3.32), we have

$$a_t + 4ik^3 a - Ca = \frac{D}{a} \int_0^x \exp(2ikx)dx, \tag{3.45}$$

where the left-hand side is entirely a function of t and the right-hand side contains a function of x as a factor. Therefore (3.45) can hold only if

$$D = 0 \tag{3.46}$$

and

$$a_t - (C - 4ik^3)a = 0, \tag{3.47}$$

which contains the unknown function $C(t)$ of t.

The asymptotic behaviour of this plane wave, when $x \to \infty$, is given by

$$\psi \sim \exp(-ikx) + b(k,t)\exp(ikx), \tag{3.48}$$

where the incident wave has been taken with amplitude unity and b is the reflection coefficient.

Substituting (3.48) in (3.36), we have

$$\exp(ikx)[b_t - 4ik^3 b - Cb] + \exp(-ikx)[4ik^3 - C] = 0$$

which implies that

$$C = 4ik^3 \tag{3.49}$$

and

$$b_t - 4ik^3 b - Cb = 0. \tag{3.50}$$

Substituting the value of C from (3.49) in (3.47) and (3.50) and integrating, we have

$$a(k,t) = a(k,0) \tag{3.51}$$

and

$$b(k,t) = b(k,0)\exp(8ik^3 t). \tag{3.52}$$

Eqns (3.43), (3.51), and (3.52) determine the evolution of the *scattering parameters* c_m, a, and b with respect to t in terms of their values at $t = 0$, which can be determined by solving the Schrödinger equation with potential $u_0(x)$, the prescribed initial value for the solution of the KdV equation. This completes the solution of the direct scattering problem.

3.5. Inverse scattering problem

We shall now solve the inverse scattering problem to determine the potential $u(x;t)$ from the knowledge of the scattering data. In the present problem this potential is the solution of the KdV equation.

We shall follow the procedure indicated by Gel'fand and Levitan (1955) and Kay and Moses (1956) and further developed by Gardner, Greene, Kruskal, and Miura (1974).

These authors show that the desired solution of the KdV equation is given by

$$u(x,t) = -2\frac{\mathrm{d}}{\mathrm{d}x}K(x,x), \tag{3.53}$$

where $K(x,y)$ satisfies the Gel'fand–Levitan integral equation

$$K(x,y) + B(x+y) + \int_{-\infty}^{\infty} B(y+z)K(x,z)\mathrm{d}z = 0 \tag{3.54}$$

and the kernel B is given by

$$B(\xi) = \sum_{m=1}^{N} c_m^2(\kappa_m,t)\exp(-\kappa_m\xi) + \frac{1}{2\pi}\int_{-\infty}^{\infty} b(k,t)\exp(ik\xi)\mathrm{d}k, \tag{3.55}$$

assuming that there are N discrete (nondegenerate) eigenvalues of the Schrödinger equation.

In eqn (3.55) the first term on the right-hand side represents the contribution of the discrete part of the spectrum, while the second term represents the contribution of the continuous part of the spectrum. With the help of eqns (3.43), (3.51) and (3.52), we can show t-dependence explicitly in eqn (3.55)

$$B(\xi) = \sum_{m=1}^{N} c_m^2(\kappa_m,0)\exp(8\kappa_m^3 t - \kappa_m\xi)$$

$$+ \frac{1}{2\pi}\int_{-\infty}^{\infty} b(k,0)\exp[i(8k^3 t + k\xi)]\mathrm{d}k. \tag{3.56}$$

Therefore, before we are able to solve the Gel'fand–Levitan integral equation to determine $K(x,y)$, we must determine $c_m(\kappa_m,0), a(k,0)$, and $b(k,0)$ by solving the direct scattering problem of the Schrödinger equation taking the initial value $u_0(x)$ as the potential.

From the foregoing discussion, we arrive at the following method of solving the KdV equation:

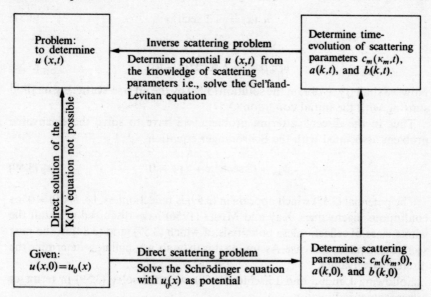

The major advantage of this method is that the solution of the nonlinear KdV equation is reduced to the solution of a linear second-order differential equation, i.e. the Schrödinger equation and a linear integral equation, i.e. the Gel'fand–Levitan equation. Moreover, in eqn (3.54) the variables t and x appear only parametrically, so that it is essentially an integral equation in a single variable y. If the potential $u_0(x)$ is reflectionless, a major simplification takes place because the second term in (3.56) gets eliminated. In fact, the initial conditions $u_0(x)$, that we shall consider in the next section, prove to be reflectionless potentials for the Schrödinger equation.

3.6.   Soliton solution of the KdV equation

In this section, we shall deal with simple cases of one soliton and two soliton solutions of the KdV equation to illustrate the inverse-scattering method. This discussion will indicate that corresponding to one discrete eigenvalue of the Schrödinger equation, there exists only one soliton solution and vice-versa. We shall find the same case in the next section in which we consider the general case of N solitons.

3.6.1.   *One-soliton solution:*

We are discussing this case simply to check the validity of our method. The solution (3.6) of eqn (3.3) corresponds to the initial condition

$$u_0(x) = -\tfrac{1}{2}a^2 \operatorname{sech}^2(\tfrac{1}{2}ax).$$

For sake of definiteness, we shall take $a = 2$ so that the steady solution of the KdV equation (3.3) corresponding to the initial condition

$$u_0(x) = -2 \operatorname{sech}^2 x \tag{3.57}$$

is

$$u(x, t) = -2 \operatorname{sech}^2(x - 4t). \tag{3.58}$$

Now we shall try to obtain the solution (3.58) by the inverse-scattering method starting with the initial condition (3.57).

Thus in the direct-scattering problem, we have to solve the eigenvalue problem associated with the Schrödinger equation:

$$\psi_{xx} + (2 \operatorname{sech}^2 x + \lambda)\psi = 0. \tag{3.59}$$

The potential (3.57) which appears in (3.59) is reflectionless, i.e. $b(k, 0) = 0$ for continuous eigenvalues. Kay and Moses (1956) have discussed in detail the whole class of reflectionless potentials of which (3.57) (and (3.84) in the next section) is a particular case. Assuming this to be so, we shall now determine the discrete eigenvalues.

Following Landau and Lifschitz (1958), we shall solve (3.59) in terms of hypergeometric functions.

We first substitute

$$\psi = \omega \operatorname{sech}^s x \tag{3.60}$$

in (3.59) to get

$$\omega_{xx} - 2s(\tanh x)\omega_x + \omega\{(2 - s - s^2)\operatorname{sech}^2 x + \lambda + s^2\} = 0. \tag{3.61}$$

If we now want the coefficient of ω to be independent of x, we must choose

$$s^2 + s - 2 = 0$$

i.e.,

$$s = 1, -2.$$

We choose

$$s = 1. \tag{3.62}$$

With this choice of s, (3.61) reduces to

$$\omega_{xx} - 2(\tanh x)\omega_x + (\lambda + 1) = 0 \tag{3.63}$$

where

$$\psi = \omega \operatorname{sech} x. \tag{3.64}$$

In order to reduce (3.63) to the standard form of the hypergeometric equation

$$x(1 - x)y_{xx} + [\gamma - (\alpha + \beta + 1)x]y_x - \alpha\beta y = 0, \tag{3.65}$$

we substitute

$$\xi = \sinh^2 x. \tag{3.66}$$

Thus we have the following equation determining ω

$$\xi(1 + \xi)\omega_{\xi\xi} + \tfrac{1}{2}\omega_\xi + \tfrac{1}{4}(1 - \kappa^2)\omega = 0, \tag{3.67}$$

where

$$\omega = (1 + \xi)^{1/2}\psi, \tag{3.68}$$

on setting formally

$$\lambda = -\kappa^2. \tag{3.69}$$

Now $u_0(x) = \dfrac{-2}{\cosh^2 x}$ is an even function of x, therefore the solution ψ of the Schrödinger equation (3.59) can be even or odd in x. Moreover, ξ defined by (3.66) is an even function of x but $\sqrt{\xi}$ is an odd function of x. We note that, from eqn (3.66),

$$\xi \to \infty \quad \text{as} \quad x \to \pm \infty.$$

Therefore, since

$$\psi \to 0 \text{ as } |x| \to \infty \text{ (i.e. as } \xi \to \infty), \frac{\omega}{(1 + \xi)^{1/2}} \to 0 \text{ as } \xi \to \infty.$$

The even and the odd (in x) integrals of (3.67) are given by

$$\omega_1 = F\left(-\frac{1}{2} + \frac{\kappa}{2}, -\frac{1}{2} - \frac{\kappa}{2}; \frac{1}{2}; -\xi\right) \tag{3.70}$$

and

$$\omega_2 = \sqrt{\xi}\, F\left(\frac{\kappa}{2}, -\frac{\kappa}{2}; \frac{3}{2}; -\xi\right). \tag{3.71}$$

For ω to be bounded at the singular point $\xi = -1$ of (3.67), (3.70) and (3.71) should reduce to polynomials. Further, we have

$$\psi_1 = \frac{F\left(-\dfrac{1}{2} + \dfrac{\kappa}{2}, -\dfrac{1}{2} - \dfrac{\kappa}{2}; \dfrac{1}{2}; -\xi\right)}{(1 + \xi)^{1/2}} \tag{3.72}$$

and

$$\psi_2 = \left(\frac{\xi}{1 + \xi}\right)^{1/2} F\left(\frac{\kappa}{2}, -\frac{\kappa}{2}; \frac{3}{2}; -\xi\right). \tag{3.73}$$

The values of κ which make ω_1 and ω_2 polynomials and give ψ_1 and ψ_2 which tend to zero as ξ tend to infinity, correspond to discrete eigenvalues.

Since $\left(\dfrac{\xi}{1+\xi}\right)^{1/2} \to 1$, as $\xi \to \infty$, eqn (3.73) is not admissible even if we take $\kappa = 0$. Eqn (3.72) will give us admissible solution only when

$$-\frac{1}{2}+\frac{\kappa}{2}=0 \quad \text{or} \quad \kappa=1 \quad \text{or} \quad \lambda=-1. \tag{3.74}$$

Thus the Schrödinger equation has only one eigenvalue $\kappa=1$ which is discrete.

The eigenfunction corresponding to the reflectionless potential (3.57) is

$$\psi = \frac{\alpha}{(1+\xi)^{1/2}} = \frac{\alpha}{\cosh x},$$

where we shall determine α from the normality condition of ψ:

$$\alpha^2 \int_{-\infty}^{\infty} \operatorname{sech}^2 x \, dx = 1$$

or

$$\alpha = \frac{1}{\sqrt{2}}$$

and the normalized eigenfunction is

$$\psi = \frac{1}{\sqrt{2}} \operatorname{sech} x. \tag{3.75}$$

From eqn (3.10), when $\kappa=1$,

$$c(0) = \lim_{x \to \infty} \psi(x) \exp x = \sqrt{2}, \tag{3.76}$$

on using eqns (3.74) and (3.75).

Since the potential (3.57) is reflectionless

$$b(k,0)=0, \qquad a(k,0)=1.$$

Therefore, from eqns (3.43) and (3.56), we have

$$c(t) = c(0) \exp(4t) = \sqrt{2} \exp(4t) \tag{3.77}$$

and

$$B(\xi,t) = 2\exp(8t - \xi) \tag{3.78}$$

so that the Gel'fand–Levitan integral equation reduces to

$$K(x,y) + 2\exp(8t - x - y) + 2\exp(8t - y) \int_{x}^{\infty} \exp(-z)K(x,z)dz = 0. \tag{3.79}$$

Considering the dependence of (3.79) on y, it appears reasonable to assume

$$K(x,y) = L(x)\exp(-y). \tag{3.80}$$

This assumption removes the y-dependence completely from (3.79) and the resulting equation yields the following expression for $L(x)$:

$$L(x) = \frac{-2\exp(x)}{1 + \exp(2x - 8t)} \tag{3.81}$$

so that the solution of the Gel'fand–Levitan equation is

$$K(x,y) = \frac{-2\exp(x - y)}{1 + \exp(2x - 8t)} \tag{3.82}$$

and the solution of the KdV equation satisfying the initial condition (3.57) is

$$u(x,t) = -2\frac{d}{dx}\left\{\frac{-2}{1 + \exp(2x - 8t)}\right\}$$

$$= -2\,\text{sech}^2(x - 4t) \tag{3.83}$$

which is the same as the solitary wave solution (3.58).

This simple example explains the working of the inverse scattering method and gives us confidence in this method. In passing, we note that corresponding to one eigenvalue $\kappa = 1$, we have just one solitary wave, as remarked earlier.

3.6.2. Two-soliton solution

Let us consider the following initial value

$$u_0(x) = -6\,\text{sech}^2 x \tag{3.84}$$

in which the amplitude and width of the wave do not match according to the solitary-wave solution (3.6) for $t = 0$.

Here we have to solve the following eigenvalue problem for the Schrödinger equation:

$$\psi_{xx} + (\lambda + 6\,\text{sech}^2 x)\psi = 0. \tag{3.85}$$

The potential $u_0(x)$ is again reflectionless. Let us try to find out the discrete eigenvalues. On substituting (3.60) in (3.59) and choosing s such that the coefficient of ω is independent of x, we have

$$s = 2, -3. \tag{3.86}$$

When we take $s = 2$ the following equation determines ω

$$\omega_{xx} - 4(\tanh x)\omega_x + (4 + \lambda)\omega = 0.$$

Now changing the independent variable to ξ as defined in (3.66), the above equation reduces to

$$\xi(1 + \xi)\omega_{\xi\xi} + (\tfrac{1}{2} - \xi)\omega_\xi + \tfrac{1}{4}(4 - \kappa^2)\omega = 0, \tag{3.87}$$

where we have formally written $-\kappa^2$ for λ.

Here also (3.84) is an even function of x, so that we shall write even and odd solutions of (3.87)

$$\omega_1 = F\left(-1 + \frac{\kappa}{2}, -1 - \frac{\kappa}{2}; \frac{1}{2}; -\xi\right) \tag{3.88}$$

and

$$\omega_2 = \sqrt{\xi} F\left(-\frac{1}{2} + \frac{\kappa}{2}, -\frac{1}{2} - \frac{\kappa}{2}; \frac{3}{2}; -\xi\right). \tag{3.89}$$

Therefore, the even and odd solutions of (3.85) are

$$\psi_1 = \frac{1}{\cosh^2 x} F\left(-1 + \frac{\kappa}{2}, -1 - \frac{\kappa}{2}; \frac{1}{2}; -\sinh^2 x\right) \tag{3.90}$$

and

$$\psi_2 = \frac{\sinh x}{\cosh^2 x} F\left(-\frac{1}{2} + \frac{\kappa}{2}, -\frac{1}{2} - \frac{\kappa}{2}; \frac{3}{2}; -\sinh^2 x\right). \tag{3.91}$$

Now ψ_1 and $\psi_2 \to 0$ as $|x| \to \infty$. These boundary conditions can be satisfied only when the hypergeometric functions in (3.90) and (3.91) reduce to polynomials. Eqn (3.90) yields admissible solution only when

$$\kappa = 2(1 - n) > 0,$$

where
$$n = 0, 1, 2, \ldots$$
This gives only one value $\kappa = 2$ corresponding to $n = 0$ for which

$$\psi_1 = \frac{1}{\cosh^2 x}, \quad \kappa = 2. \tag{3.92}$$

Eqn (3.91) reduces to a polynomial, when

$$\kappa = 1 - 2n > 0, \qquad n = 0, 1, 2, \ldots.$$

Here again we get only one value of κ, namely $\kappa = 1$ corresponding to $n = 0$ which gives the admissible solution

$$\psi_2 = \frac{\sinh x}{\cosh^2 x}, \quad \kappa = 1. \tag{3.93}$$

The normalized eigenfunctions corresponding to eigenvalues $\kappa = 2$ and $\kappa = 1$

are respectively

$$\psi_1 = \frac{\sqrt{3}}{2}\operatorname{sech}^2 x, \quad \kappa_1 = 2 \tag{3.92'}$$

and

$$\psi_2 = \sqrt{\frac{3}{2}}\operatorname{sech}^2 x \sinh x, \quad \kappa_2 = 1. \tag{3.93'}$$

We can immediately determine the values of $c_1(0)$ and $c_2(0)$ as in (3.76)

$$c_1(0) = \lim_{x \to \infty} \exp(2x)\frac{\sqrt{3}}{2}\frac{1}{\cosh^2 x} = 2\sqrt{3} \tag{3.94}$$

and

$$c_2(0) = \lim_{x \to \infty} \exp(x)\sqrt{\frac{3}{2}}\frac{\sinh x}{\cosh^2 x} = \sqrt{6}. \tag{3.95}$$

Therefore, from eqn (3.43), we have

$$c_1(t) = 2\sqrt{3}\exp(32t) \tag{3.96}$$

and $\quad\quad c_2(t) = \sqrt{6}\exp(4t).$ (3.97)

Here the kernel $B(\xi)$ is given by

$$B(\xi) = 12\exp(64t - 2\xi) + 6\exp(8t - \xi) \tag{3.98}$$

so that the Gel'fand–Levitan integral equation takes the following form:

$$K(x,y) + \{12\exp(64t - 2x - 2y) + 6\exp(8t - x - y)\}$$

$$+ \int_x^\infty [\{12\exp(64t - 2y - 2z) + 6\exp(8t - y - z)\}K(x,z)]dz = 0. \tag{3.99}$$

To remove the y-dependence in the above equation, we assume

$$K(x,y) = C_1 L_1(x) \exp(-2y) + C_2 L_2(x) \exp(-y), \tag{3.100}$$

where

$$C_1 = 12\exp(64t) \quad \text{and} \quad C_2 = 6\exp(8t). \tag{3.101}$$

The above substitution in (3.99) yields the following two equations for determining $L_1(x)$ and $L_2(x)$ on equating the coefficients of $\exp(-2y)$ and $\exp(-y)$ separately equal to zero and performing the necessary integration:

$$\left\{1 + \frac{C_1}{4}\exp(-4x)\right\}L_1 + \frac{C_2}{3}\exp(-3x)L_2 + \exp(-2x) = 0$$

and

$$\frac{C_1}{3}\exp(-3x)L_1 + \left\{1 + \frac{C_2}{2}\exp(-2x)\right\}L_2 + \exp(-x) = 0.$$

Solving these equations, we have

$$L_1 = -\frac{1}{D}\left\{\exp(-2x) + \frac{1}{6}C_2\exp(-4x)\right\}$$

and

$$L_2 = \frac{1}{D}\left\{-\exp(-x) + \frac{1}{12}C_1\exp(-5x)\right\},$$

where $$D = 1 + \frac{C_1}{4}\exp(-4x) + \frac{C_2}{2}\exp(-2x) + \frac{C_1 C_2}{72}\exp(-6x).$$

Substituting these values of L_1 and L_2 in eqn (3.100) and identifying y with x, we have

$$K(x,x) = -6\frac{\exp(8t - 2x) + 2\exp(64t - 4x) + \exp(72t - 6x)}{1 + 3\exp(8t - 2x) + 3\exp(64t - 4x) + \exp(72t - 6x)}$$

which yields the following solutions for the KdV equation satisfying the initial condition (3.84)

$$u(x,t) = -12\frac{3 + 4\cosh(8t - 2x) + \cosh(64t - 4x)}{[\cosh(36t - 3x) + 3\cosh(28t - x)]^2}. \tag{3.102}$$

We shall now consider the asymptotic behaviour of (3.102) as $x \to \pm\infty$ to study its nature. Our aim in studying this asymptotic behaviour is to see if it represents hidden in it the soliton solutions corresponding to eigenvalues $\kappa_1 = 2$ and $\kappa_2 = 1$.

We first examine the eigenvalue $\kappa_1 = 2$. For this, we set

$$\xi = x - 4\kappa_1^2 t = x - 16t \tag{3.103}$$

in (3.102) and take the limit as $t \to \pm\infty$ keeping ξ fixed. We can easily show that

$$\lim_{\substack{t \to -\infty \\ \xi \text{ fixed}}} u(x,t) = -96\frac{\exp(4\xi)}{\{1 + 3\exp(4\xi)\}^2}$$

$$= -8\operatorname{sech}^2 2(\xi - \xi_1)$$

$$= -8\operatorname{sech}^2 \{2(x - 16t - \xi_1)\}, \tag{3.104}$$

where

$$\exp(-4\xi_1) = 3 \tag{3.105}$$

and

$$\lim_{\substack{t \to +\infty \\ \xi \text{ fixed}}} u(x,t) = -96 \frac{\exp(-4\xi)}{\{1 + 3\exp(-4\xi)\}^2}$$

$$= -8 \operatorname{sech}^2 2(\xi - \xi_1')$$

$$= -8 \operatorname{sech}^2 \{2(x - 16t - \xi_1')\}, \tag{3.106}$$

where

$$\exp(4\xi_1') = 3. \tag{3.107}$$

Similarly, if we set

$$\xi = x - 4\kappa_2^2 t = x - 4t \tag{3.108}$$

and take the limit of (3.102) when $t \to \pm\infty$ keeping ξ fixed, we get

$$\lim_{\substack{t \to -\infty \\ \xi \text{ fixed}}} u(x,t) = -24 \frac{\exp(-2\xi)}{\{1 + 3\exp(-2\xi)\}^2}$$

$$= -2 \operatorname{sech}^2(\xi - \xi_2)$$

$$= -2 \operatorname{sech}^2(x - 4t - \xi_2), \tag{3.109}$$

where

$$\exp(2\xi_2) = 3 \tag{3.110}$$

and

$$\lim_{\substack{t \to \infty \\ \xi \text{ fixed}}} u(x,t) = -24 \frac{\exp(2\xi)}{\{1 + 3\exp(2\xi)\}^2}$$

$$= -2 \operatorname{sech}^2(\xi - \xi_2')$$

$$= -2 \operatorname{sech}^2(x - 4t - \xi_2'), \tag{3.111}$$

where

$$\exp(-2\xi_2') = 3. \tag{3.112}$$

Now we recognize that eqns (3.104) and (3.106) represent the same soliton travelling from $x = -\infty$ to $x = \infty$ except that they differ in phase. Similarly, eqns (3.109) and (3.111) represent the same soliton travelling from $x = -\infty$ to $x = \infty$; they also differ in phase. Thus we conclude that to each discrete eigenvalue of the Schrödinger equation (3.85), there corresponds a soliton solution. In the present case, therefore, eqn (3.102) represents a *double* wave solution which breaks into two solitons both as $t \to -\infty$ and as $t \to \infty$ and the effect of nonlinear interaction between them according to the KdV equation

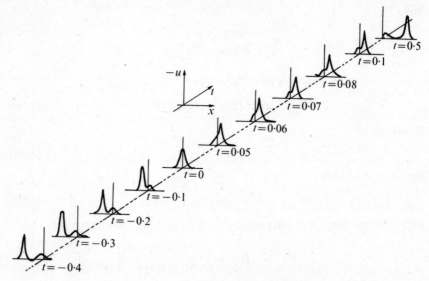

Fig. 3.1. Interaction of two solitons.

simply displaces their relative positions with respect to their positions where they would have been in the absence of the interaction.

Fig. 3.1 represents interaction of two disparate solitons with different velocities 16 and 4 and amplitudes 8 and 2 respectively starting at $t = -\infty$, the bigger following the smaller. The actual interaction takes place mainly from $t = -0.4$ to $t = 0.4$. At first the bigger soliton begins to swallow the smaller one; at $t = 0$ they both combine and form a single double wave, then the bigger soliton emerges out leaving the smaller behind it. At $t = 0.5$, they again appear as separate solitons. This soliton behaviour is very surprising as mentioned earlier, because it is exactly similar to linear interaction except for the phase difference.

3.6.3. N-soliton solution

In studying the N-soliton solution of the KdV equation we shall essentially follow the method indicated in §3.6.2 with necessary modification.

We *assume* that the potential $u_0(x)$ is reflectionless and the Schrödinger equation has N discrete eigenvalues κ_m, $m = 1, 2, \ldots, N$. From eqn (3.55), the kernel $B(\xi)$ here is

$$B(\xi) = \sum_{m=1}^{N} c_m^2 \exp(-\kappa_m \xi) \tag{3.113}$$

where

$$c_m^2 = c_m^2(0) \exp(8\kappa_m^3 t) \tag{3.114}$$

and the Gel'fand–Levitan integral equation takes the form

$$K(x,y) + \sum_{m=1}^{N} c_m^2 \exp\{-\kappa_m(x+y)\}$$

$$+ \sum_{m=1}^{N} c_m^2 \exp(-\kappa_m y) \int_x^\infty \exp(-\kappa_m z) K(x,z) \mathrm{d}z = 0. \tag{3.115}$$

In order to remove the y-dependence from (3.115), we must take $K(x,y)$ in the following form

$$K(x,y) = -\sum_{m=1}^{N} c_m \psi_m(x) \exp(-\kappa_m y), \tag{3.116}$$

where $\psi_m(x)$ are unknown functions and the factors c_m have been introduced so that ψ_m turn out to be normalized eigenfunctions of the Schrödinger equation

$$\psi_{xx} - (\kappa_m^2 + u)\psi = 0. \tag{3.117}$$

We now substitute (3.116) in (3.115) and separately equate the coefficients of $\exp(-\kappa_m y)$ to get the following set of N linear equations for determining the unknown ψ functions:

$$\psi_m(x) + \sum_{n=1}^{N} c_m c_n \psi_n(x) \frac{\exp\{-(\kappa_m + \kappa_n)x\}}{\kappa_m + \kappa_n} = c_m \exp(-\kappa_m x), m = 1,2,\ldots,N. \tag{3.118}$$

We can write the above set of equations in the matrix notation as

$$(I + C)\psi = E, \tag{3.119}$$

where I is the unit matrix of order N,

$$C \equiv [c_{mn}] = \left[c_m c_n \frac{\exp\{-(\kappa_m + \kappa_n)x\}}{\kappa_m + \kappa_n} \right] \tag{3.120}$$

is a $N \times N$ matrix, and

$$\psi = \begin{bmatrix} \psi_1 \\ \psi_2 \\ . \\ . \\ . \\ \psi_N \end{bmatrix} \quad \text{and } E = \begin{bmatrix} c_1 \exp(-\kappa_1 x) \\ c_2 \exp(-\kappa_2 x) \\ . \\ . \\ . \\ c_N \exp(-\kappa_N x) \end{bmatrix} \tag{3.121}$$

are the column matrices.

A sufficient condition that (3.119) has a unique solution is that C is positive definite, which, indeed, is the case as seen below. The quadratic form corresponding to the square matrix C is

$$\sum_{m=1}^{N} \sum_{n=1}^{N} c_m c_n \frac{\exp\{-(\kappa_m + \kappa_n)x\}}{\kappa_m + \kappa_n} X_m X_n$$

$$= \int_x^{\infty} dz \left[\sum_{m=1}^{N} c_m \exp(-\kappa_m z) X_m \right]^2,$$

which is certainly positive definite. This completes the proof of the above statement.

Now we can easily prove that

$$0 < \det C = \left(\prod_{m=1}^{N} c_m^2 \right) \exp \left\{ -2 \left(\sum_{m=1}^{N} \kappa_m \right) x \right\} \det \left(\frac{1}{\kappa_m + \kappa_n} \right) \quad (3.122)$$

so that

$$\det \left[\frac{1}{\kappa_m + \kappa_n} \right] > 0. \quad (3.123)$$

From (3.122), it is clear that we can write

$$\det C = \alpha \exp(-\beta x), \quad (3.124)$$

where both α and β are positive.

Let Q_{mn} denote the cofactors of the coefficient matrix $I + C$ in (3.119), then expanding in terms of the nth column, we have

$$\Delta \equiv \det(I + C) = \sum_{m=1}^{N} \left(\delta_{mn} + c_m c_n \frac{\exp\{-(\kappa_m + \kappa_n)x\}}{\kappa_m + \kappa_n} \right) Q_{mn}, \quad (3.125)$$

where δ_{mn} are the Kronecker deltas. Then Cramer's rule gives us

$$\psi_m = \frac{1}{\Delta} \sum_{n=1}^{N} c_n \exp(-\kappa_n x) Q_{mn}. \quad (3.126)$$

Therefore, we have, on identifying y with x,

$$K(x,x) = -\sum_{m=1}^{N} c_m \psi_m(x) \exp(-\kappa_m x)$$

$$= -\frac{1}{\Delta} \sum_{m=1}^{N} \sum_{n=1}^{N} c_m c_n \exp\{-(\kappa_m + \kappa_n)x\} \cdot Q_{mn}$$

$$= \frac{1}{\Delta} \frac{d}{dx} \Delta = \frac{d}{dx} \ln \Delta, \quad (3.127)$$

and then the desired solution of the KdV equation is

$$u(x,t) = -2\frac{d}{dx}K(x,x) = -2\frac{d^2}{dx^2}\ln\Delta \tag{3.128}$$

$$= -2\frac{d^2}{dx^2}\ln\{\det(I+C)\}. \tag{3.129}$$

We shall now show that $\psi_m(x)$ in (3.116) are normalized eigenfunctions corresponding to the eigenvalues of the Schrödinger equation

$$L_m[\psi_m] = 0, \qquad L_m \equiv \frac{d^2}{dx^2} - [\kappa_m^2 + u(x)], \tag{3.130}$$

where L_m is a linear differential operator. Operating by L_m on (3.118) we have, after considerable simplification,

$$L_m[\psi_m] + \sum_{n=1}^{N}\frac{c_m c_n}{\kappa_m + \kappa_n}\exp\{-(\kappa_m + \kappa_n)x\}L_n[\psi_n] = 0$$

or

$$(I+C)L[\psi] = 0, \tag{3.131}$$

where $L[\psi]$ is the column vector $\begin{bmatrix} L_1[\psi_1] \\ L_2[\psi_2] \\ \cdot \\ \cdot \\ \cdot \\ L_N[\psi_N] \end{bmatrix}$,

Since $I + C$ is nonsingular, (3.131) admits only the trivial solution

$$L[\psi] = 0.$$

Therefore ψ_m are the eigenfunctions of the Schrödinger equation corresponding to eigenvalues κ_m.

If we write (3.118) in the form

$$\psi_m\exp(\kappa_m x) + \sum_{n=1}^{N}\frac{c_m c_n}{\kappa_m + \kappa_n}\cdot\{\psi_n\exp(\kappa_n x)\}\exp(-2\kappa_n x) = c_m$$

and proceed to limit as $x \to \infty$, we have

$$\lim_{x \to \infty}\psi_m\exp(\kappa_m x) = c_m$$

so that ψ_m are the normalized eigenfunctions.

3.7. Soliton interaction

We shall now study the behaviour of the solution (3.129) as $t \to \pm\infty$ to prove that it does contain N soliton solutions embedded in it.

Let us introduce the following notation:

$$u(x,t) = -2\frac{d}{dx}K(x,x) = 2\frac{d}{dx}\sum_{m=1}^{N} c_m\psi_m(x)\exp(-\kappa_m x)$$

$$= 2\frac{d}{dx}\sum_{m=1}^{N} k_m(x)$$

$$= 2\sum_{m=1}^{N} k'_m(x), \tag{3.132}$$

where

$$\left.\begin{array}{l} k_m(x) = c_m\psi_m(x)\exp(-\kappa_m x) \\[2mm] \psi_m(x) = \dfrac{k_m}{c_m}\exp(\kappa_m x). \end{array}\right\} \tag{3.133}$$

or

To evaluate k_m, we write (3.118) in terms of k_m

$$c_m^{-2}\exp(2\kappa_m x)k_m + \sum_{n=1}^{N} \frac{k_n}{\kappa_m + \kappa_n} = 1, \quad m = 1, 2, \ldots, N. \tag{3.134}$$

Differentiating (3.134) with respect to x, we have

$$c_m^{-2}\exp(2\kappa_m x)k'_m + \sum_{n=1}^{N} \frac{k'_n}{\kappa_m + \kappa_n} = -2\kappa_m c_m^{-2}\exp(2\kappa_m x)k_m,$$

$$m = 1, 2, \ldots, N. \tag{3.135}$$

We note that we could solve (3.134) and get k_m, and then evaluate k'_m which we need in (3.132). However, we shall find it more elegant and convenient to deal with (3.135). We now transform eqns (3.134) and (3.135) to a moving-coordinate system moving with the natural free speed of one of the solitons, defined by

$$\xi = x - 4\kappa_p^2 t, \quad p = 1, 2, \ldots, \text{or } N. \tag{3.136}$$

Eqns (3.134) and (3.135) thus reduce to

$$C_m(\xi)\exp\{-8\kappa_m(\kappa_m^2 - \kappa_p^2)t\} \cdot k_m + \sum_{n=1}^{N} \frac{k_n}{\kappa_m + \kappa_n} = 1 \tag{3.137}$$

$$C_m(\xi)\exp[-8\kappa_m(\kappa_m^2 - \kappa_p^2)t] \cdot k'_m + \sum_{n=1}^{N} \frac{k'_n}{\kappa_m + \kappa_n}$$

$$= -2\kappa_m k_n C_m(\xi)\exp\{-8\kappa_m(\kappa_m^2 - \kappa_p^2)t\}, \tag{3.138}$$

where

$$C_m(\xi) = c_m^{-2}(0)\exp(2\kappa_m \xi). \tag{3.139}$$

When $m = p$, the above equations yield

$$C_p(\xi)k_p + \sum_{n=1}^{N} \frac{k_n}{\kappa_m + \kappa_n} = 1 \tag{3.140}$$

and

$$C_p(\xi)k_p' + \sum_{n=1}^{N} \frac{k_n'}{\kappa_m + \kappa_n} = -2\kappa_p k_p C_p(\xi). \tag{3.141}$$

For our further work, we order the eigenvalues as

$$\kappa_1 > \kappa_2 > \ldots > \kappa_N. \tag{3.142}$$

3.7.1. *Asymptotic behaviour as* $t \to \infty$

When $t \to \infty$, (3.137) reduces to

$$\sum_{n=1}^{N} \frac{k_n}{\kappa_m + \kappa_n} = 1, \qquad m = 1, 2, \ldots, p-1 \tag{3.143}$$

$$C_p(\xi)k_p + \sum_{n=1}^{N} \frac{k_n}{\kappa_m + \kappa_n} = 1, \quad m = p \tag{3.144}$$

and

$$k_m = 0, \qquad m = p+1, \ldots, N. \tag{3.145}$$

In view of (3.145), we can write (3.143) and (3.144) in the combined form as

$$\sum_{n=1}^{p} \frac{k_n}{\kappa_m + \kappa_n} = 1 - \delta_{mp} C_p(\xi)k_p, \quad m = 1, 2, \ldots, p,$$

and

$$k_m = 0 \quad \text{for } m = p+1, p+2, \ldots, N. \tag{3.146}$$

Similarly, (3.138) and (3.141) yield

$$\sum_{n=1}^{p} \frac{k_n'}{\kappa_m + \kappa_n} = -C_p(\xi)\delta_{mn}(2\kappa_p k_p + k_p'), \; m = 1, 2, \ldots, p,$$

and $\quad k_m' = -2\kappa_m k_m = 0, \quad \text{for } m = p+1, \quad p+2, \ldots, N. \tag{3.147}$

For solving (3.146) and (3.147), we define the matrices

$$K_p \equiv \left[\frac{1}{\kappa_m + \kappa_n} \right], \quad m, n = 1, 2, \ldots, p. \tag{3.148}$$

which have positive determinants for all $p = 1, 2, \ldots, N$.

Let L_p denote the matrix obtained from K_p by replacing its last column by a column in which all elements are 1.

From (3.146) and (3.147) by Cramer's rule, we have

$$k_m \det K_p = \sum_{n=1}^{p} K_{mn} - C_p(\xi) K_{pm} k_p, m = 1, 2, \ldots, p \qquad (3.149)$$

and

$$k'_m \det K_p = - C_p(\xi) K_{pm} (2\kappa_p k_p + k'_p), \quad m = 1, 2, \ldots, p, \qquad (3.150)$$

where K_{mn} denotes the cofactor of $\dfrac{1}{\kappa_m + \kappa_n}$ in K_p. Taking $m = p$ and solving for k_p and k'_p, we have

$$k_p = \frac{\det L_p}{\det K_p + C_p(\xi) \det K_{p-1}} \qquad (3.151)$$

and

$$k'_p = - \frac{2 C_p(\xi) \kappa_p k_p \det K_{p-1}}{\det K_p + C_p(\xi) \det K_{p-1}}. \qquad (3.152)$$

Summing (3.150) and using (3.151) and (3.152) we have

$$\lim_{\substack{t \to \infty \\ \xi \text{ fixed}}} \sum_{m=1}^{p} k'_m = \frac{-2\kappa_p C_p(\xi)}{\left[\dfrac{\det K_p}{\det L_p} + C_p(\xi) \dfrac{\det K_{p-1}}{\det L_p} \right]^2}. \qquad (3.153)$$

We can easily evaluate the value of $\det K_p$ by subtracting the last column ($n = p$) from each of the other columns and taking out common factors from the rows and the columns:

$$\det K_p = \frac{\displaystyle\prod_{m=1}^{p-1} (\kappa_p - \kappa_m)}{\displaystyle\prod_{m=1}^{p} (\kappa_p + \kappa_m)} \det L_p. \qquad (3.154)$$

Similarly, subtracting the last row ($n = p$) of $\det L_p$ from each of the other rows, taking out the common factors and then expanding in terms of the last column (consisting of all zeros except the last which is 1), we obtain

$$\det L_p = \frac{\displaystyle\prod_{m=1}^{p-1} (\kappa_p - \kappa_m)}{\displaystyle\prod_{m=1}^{p-1} (\kappa_p - \kappa_m)} \det K_{p-1}. \qquad (3.155)$$

Using (3.139), (3.154) and (3.155) in (3.153), we have

$$\lim_{\substack{t \to \infty \\ \xi \text{ fixed}}} u(x,t) = \lim_{\substack{t \to \infty \\ \xi \text{ fixed}}} 2 \sum_{m=1}^{p} k'_m$$

$$= \frac{-8\kappa_p^2 \exp\{2\kappa_p(\xi - \xi_p)\}}{[1 + \exp\{2\kappa_p(\xi - \xi_p)\}]^2}$$

$$= -2\kappa_p^2 \operatorname{sech}^2[\kappa_p(\xi - \xi_p)]$$

$$= -2\kappa_p^2 \operatorname{sech}^2[\kappa_p(x - 4\kappa_p^2 t - \xi_p)], \qquad (3.156)$$

where ξ_p is given by

$$\exp\{2\kappa_p \xi_p\} = \frac{c_p^2(0)}{2\kappa_p} \prod_{m=1}^{p-1} \left(\frac{\kappa_p - \kappa_m}{\kappa_p + \kappa_m}\right)^2. \qquad (3.157)$$

(3.156) is identical in form to a solitary wave with amplitude $2\kappa_p^2$ moving in the positive direction of the x-axis with a constant speed $4\kappa_p^2$.

3.7.2. *Asymptotic behaviour as* $t \to -\infty$

In this case we have the following equations corresponding to (3.146) and (3.147)

$$\sum_{n=p}^{N} \frac{k_n}{\kappa_m + \kappa_n} = 1 - C_p(\xi)\delta_{mp}k_p, \quad m = p, p+1, \ldots, N$$

$$k_m = 0, \quad m = 1, 2, \ldots, p-1 \qquad (3.158)$$

and

$$\left. \begin{aligned} \sum_{n=p}^{N} \frac{k'_n}{\kappa_m + \kappa_n} &= -C_p(\xi)\delta_{mp}(2\kappa_p k_p + k'_p) \\ & \qquad\qquad m = p, p+1, \ldots, N \\ k'_m = -2\kappa_m k_m &= 0, \qquad m = 1, 2, \ldots, p-1. \end{aligned} \right\} \qquad (3.159)$$

Eqns (3.158) and (3.159) have the same structure as (3.146) and (3.147) respectively except that the range 1 to p has been replaced by p to N. Therefore

$$\lim_{\substack{t \to -\infty \\ \xi \text{ fixed}}} u(x,t) = -2\kappa_p^2 \operatorname{sech}^2\{\kappa_p(\xi - \xi'_p)\}$$

$$= -2\kappa_p^2 \operatorname{sech}^2\{\kappa_p(x - 4\kappa_p^2 t - \xi'_p)\} \qquad (3.160)$$

where

$$\exp(2\kappa_p \xi'_p) = \frac{c_p^2(0)}{2\kappa_p} \prod_{m=p+1}^{N} \left(\frac{\kappa_p - \kappa_m}{\kappa_p + \kappa_m}\right)^2. \qquad (3.161)$$

The foregoing discussion leads to the following conclusion:

If $u(x,t)$, a solution of the KdV equation, is a reflectionless potential of the Schrödinger equation, then as $t \to \pm \infty$ each eigenvalue $\lambda_p = -\kappa_p^2$ has associated with it a soliton which approaches the solitary wave-form (3.156) for $t \to \infty$ and solitary wave-form (3.160) as $t \to -\infty$. The solitary wave has uniform speed $4\kappa_p^2$ and amplitude $2\kappa_p^2$. During its journey from $t = -\infty$ to $t = \infty$, its trajectory is shifted by

$$\xi_p - \xi_p' = \frac{1}{\kappa_p} \left\{ \sum_{m=1}^{p-1} \ln \left(\frac{\kappa_m - \kappa_p}{\kappa_m + \kappa_p} \right) - \sum_{m=p+1}^{N} \ln \left(\frac{\kappa_p - \kappa_m}{\kappa_p + \kappa_m} \right) \right\} \quad (3.162)$$

This theorem has been independently proved by Zakharov (1971), Wadati and Toda (1972), and Tanaka (1972–73).

From (3.162) we note that the total shift is just the sum of the shifts (that would be undergone) in isolated pairwise interaction with every other soliton. Zakharov (1971) has also pointed out this fact.

3.8.   Continuous eigenvalues of the Schrödinger operator

We have so far discussed only the waves corresponding to the discrete eigenvalues of the Schrödinger operator. Ablowitz and Newell (1973) have discussed the problem corresponding to the continuous eigenvalues. Taking the KdV equation in the form

$$u_t + uu_x + u_{xxx} = 0$$

they have established that, if the initial data does not give rise to discrete eigenvalues, the solution $u(x,t)$ shows that for $x \gg t^{1/3}$, the solution decays exponentially.

However, their results for $|x| = O(t^{1/3})$ and $x \ll -t^{1/3}$ appear to be incorrect. For further comments on the solution contributed by the continuous eigenvalues, reference may be made to the recent review article of Miura (1976).

To the best of our knowledge, the case when the associated spectrum consists of both discrete as well as continuous parts has not been studied.

Bibliography

ABLOWITZ, M. J. and NEWELL, A. C. The decay of the continuous spectrum for solutions of the Korteweg–de Vries equation. *J. Math. Phys.* **14**, 1277–84 (1973).

GARDNER, C. S. GREENE, J. M. KRUSKAL, M. D. and MIURA, R. M. Korteweg–de Vries equation and generalizations. VI. Methods for exact solution. *Communs pure appl. Math.* **27**, 97–133 (1974).

GEL'FAND, I. M. and LEVITAN, B. M. On the determination of a differential equation from its spectral function. *Am. math. Soc. Transl.* Series 2 **1**, 253–304 (1955).

KAY, I. and Moses, H. E. Reflectionless transmission through dielectrics and scattering potentials. *J. appl. Phys.* **27**, 1503–8 (1956).

LANDAU, L. and LIFSCHITZ, M. *Quantum mechanics non-relativistic theory.* Pergamon Press, New York (1958).

MIURA, R. M. The Korteweg–de Vries equation: a survey of results. *SIAM Rev.* **18**, 412–59 (1976).

——, GARDNER, C. S. and KRUSKAL, M. D. Korteweg–de Vries equation and generalizations II. Existence of conservation laws and constants of motion. *J. math. Phys.* **9**, 1204–9 (1968).

TANAKA, S. On the N-tuple wave solutions of the Korteweg–de Vries equation. *Publ. Res. Inst. Math. Sci. Kyoto Univ.* **8**, 419–27 (1972–73).

WADATI, M. and TODA, M. The exact N-soliton solution of the Korteweg–de Vries equation. *J. phys. Soc. Japan.* **32**, 1403–11 (1972).

ZAKHAROV, V. E. Kinetic equation for solitons. *Soviet Phys. JETP.* **33**, 538–41 (1971).

4

GENERAL EQUATION OF EVOLUTION

4.1. Introduction

The discussion of the soliton interaction presented in Chapter 3 was based on the fact that it was possible to associate the nonlinear KdV equation with the time-independent Schrödinger equation in one-dimension which is linear. The time t, which occurs in the solution $u = u(x,t)$ of the KdV equation, behaved as a parameter when $u(x,t)$ was taken as the potential in the Schrödinger equation. This technique enabled us to utilize the well-known properties of the eigenvalues and eigenfunctions of the Schrödinger equation. One newly discovered specific property on which the success of the inverse scattering method depended was that the energy eigenvalues, both discrete and continuous, of the Schrödinger operator in which the potential evolved according to the KdV equation, are time-independent. Therefore, they could be determined once for all taking the initial value $u(x,0)$ as the potential in the Schrödinger operator.

Lax (1968), in his important paper, examined if the general equation of evolution

$$u_t = K(u), \tag{4.1}$$

where K is a nonlinear operator acting on u but which does not contain the independent variables x and t explicitly, can also be studied in the similar manner. His answer was in the affirmative.

We shall now discuss Lax's method for studying general equations of evolution of the type (4.1) and then apply it to the study of the KdV equation, which we have studied earlier, to illustrate this method. It is clear that the mathematical strategy for studying a general equation (4.1) has, by necessity, to be general as we shall see in the following sections.

It is interesting to note that Lax's discussion and assumptions run parallel to those used in the study of the KdV equation. However, it is not surprising because his aim was mainly to explain the existence of the hidden solitons in an arbitrary solution.

We shall study the solutions of the KdV equation which belong to C^∞ class of functions defined over the real line: $-\infty < x < \infty$ and which, along with their derivatives of all orders, tend to zero as $|x| \to \infty$. We note that the solitary-wave solutions of the KdV equation also belong to this class of functions. We further note that we have studied the existence of solitons through the asymptotic behaviour of reflectionless solutions. In particular, we have shown that there exist a set of positive speeds $c_p = 4\kappa_p^2, p = 1, 2, \ldots, N$, and a corresponding set of phases $\xi_p^{\pm}$ such that

$$\lim_{\substack{t \to \pm \infty \\ \xi \text{ fixed}}} u(x,t) = \begin{cases} s(\xi - \xi_p^{\pm}, c) & \text{if } c = c_p \\ 0 & \text{if } c \neq c_p, \end{cases} \tag{4.2}$$

where $\xi = x - ct$ and is the moving coordinate moving with velocity c, and κ_p are the discrete eigenvalues of the Schrödinger operator with reflectionless potential $u(x,t)$. Our aim in this chapter is to show that the result (4.2) is true for an arbitary solution $u(x,t)$ of the KdV equation.

Thus c_p depends on the choice of the solution $u(x,t)$ of the KdV equation. We shall express this fact by saying that c_p is a functional of u and symbolically write

$$c_p = c_p(u). \tag{4.3}$$

We shall denote a solitary wave solution of (4.1) by

$$s = s(\xi), \qquad \xi = x - ct. \tag{4.4}$$

We introduced the concept of an integral of the KdV equation in the previous chapter and according to it the eigenspeeds c_p are invariant functionals or integrals. Mathematically, we can express this invariance as follows.

If u' and u'' are the values of u at two distinct instants of time t' and t'', then

$$c_p(u') = c_p(u''). \tag{4.5}$$

Similarly, from our discussion of soliton solutions, we conclude that the phase differences $\xi_p^+ - \xi_p^-$ are also 'integrals' of the KdV equation. We have also mentioned in the last chapter that Miura, Gardner, and Kruskal succeeded in proving the existence of an infinite sequence of integrals of the KdV equation. They also gave a method of construction of these integrals. Integrals or time-invariant functionals of the solutions of the general evolution equation (4.1) may also exist. In this chapter we shall study the properties of integrals of (4.1).

Our first task would be to associate a linear equation corresponding to the Schrödinger equation with (4.1) and then prove the time-invariance of the eigenvalues of this equation. These eigenvalues will then be the integrals of (4.1). The following important properties of the KdV equation will suggest the assumption that we should make in the study of eqn (4.1).

(a) The KdV equation always admits a solution corresponding to a given initial condition which is smooth enough and tends to zero rapidly as $|x| \to \infty$.

(b) The solutions of the KdV equation, which belong to $C^{\infty}(-\infty, \infty)$ and which, along with their x-derivatives of all order, tend to zero as $|x| \to \infty$, are *uniquely* determined by their initial values.

The proof of this uniqueness theorem is straight forward. Let $u_0(x)$ be the given initial condition. If possible, let u and v be two solutions having $u_0(x)$ as the initial condition. Then

$$u_t + uu_x + u_{xxx} = 0.$$

$$v_t + vv_x + v_{xxx} = 0.$$

Subtracting the second equation from the first and writing w for $u - v$, we have

$$w_t + uw_x + v_x w + w_{xxx} = 0.$$

Evidently, w belongs to the $C^\infty(-\infty, \infty)$ class and tends to zero along with its derivatives of all orders as $|x| \to \infty$.

Multiplying the last equation by w and integrating from $x = -\infty$ to $x = \infty$, we have

$$\frac{d}{dt} \int_{-\infty}^{\infty} \frac{1}{2} w^2 dx + \int_{-\infty}^{\infty} \left(v_x - \frac{1}{2} u_x \right) w^2 dx = 0.$$

Writing $\displaystyle \int_{-\infty}^{\infty} \frac{1}{2} w^2 dx = E(t)$ and denoting $\displaystyle \max_{\substack{-\infty < x < \infty \\ t \geq 0}} |v_x - \tfrac{1}{2} u_x|$ by m, we have

$$\frac{dE}{dt} \leq mE$$

which on integration yields

$$E(t) \leq E(0) \exp(mt).$$

Since $\quad E(0) = \displaystyle \int_{-\infty}^{\infty} \tfrac{1}{2} [u(x,0) - v(x,0)]^2 dx = 0,$

$E(t) = 0$ for all $t \geq 0$.

Hence $w \equiv 0$. This proves the uniqueness of the solution.

In view of the two properties (a) and (b), the eigenspeeds are functionals of the initial conditions.

In Chapter 3, we have discussed the KdV equation in the form

$$u_t - 6uu_x + u_{xxx} = 0$$

and the associated Schrödinger equation

$$\frac{d^2 \psi}{dx^2} + (\lambda - u)\psi = 0, \quad \lambda = -\kappa^2.$$

Consequently, the solitary-wave solution was in the form

$$s(x,t) = -2\kappa^2 \operatorname{sech}^2 \{\kappa(x - 4\kappa^2 t)\}.$$

In the present discussion we shall take the KdV equation in the form

$$u_t + uu_x + u_{xxx} = 0 \qquad (4.6)$$

and write $-\lambda$ for λ, so that the corresponding Schrödinger equation and the solitary-wave solution have to be taken in the following forms:

$$\frac{d^2\psi}{dx^2} + \left[-\lambda + \frac{1}{6}u \right]\psi = 0$$

or

$$L\psi = \lambda\psi, \quad L \equiv \frac{d^2}{dx^2} + \frac{1}{6}u, \quad \lambda = \kappa^2 \qquad (4.7)$$

and

$$s(x,t) = 3c \operatorname{sech}^2 \left\{ \frac{\sqrt{c}}{2}(x - ct) \right\},$$

where the eigenspeed $\quad c = 4\kappa^2.$ $\qquad (4.8)$

We note that the above choice has been made only for sake of convenience in analysis and does not affect the generality of the treatment.

4.2. Definitions

In this section, we shall introduce some definitions which we need for our discussion.

Let B be some space of functions, such that, for each t, a solution $u(t)$ of the equation of evolution

$$u_t = K(u) \qquad (4.9)$$

belongs to B.

We *assume* that we can associate to each function $u\varepsilon B$ a self-adjoint operator $L = \underset{u}{L}$ over some Hilbert space H, where u written below L indicates that it depends on u:

$$u \to \underset{u}{L}$$

with the following property: If the time-variation of u is governed by eqn (4.9), then the operators $L(t)$, which also change with t remain *unitarily equivalent*.

To define the term 'unitarily equivalent' we have to introduce some more definitions which are very well known, but we record them here for sake of completeness.

4.2.1. *Adjoint operator*

Let $\phi, \psi \varepsilon H$, and L be an operator defined in H, then the operator L^* is said to be adjoint to L, if

$$(L\phi, \psi) = (\phi, L^*\psi) \qquad (4.10)$$

for all ϕ and ψ for which the operator is defined, where (,) denotes an inner product in H.

4.2.2 Self-adjoint operator

An operator L is said to be self-adjoint if it is defined for every element εH and it is identical with its adjoint operator, i.e.,

$$L^* = L,$$

so that
$$(L\phi,\psi) = (\phi,L\psi) \tag{4.11}$$

for all ϕ and $\psi \varepsilon H$, if L is self-adjoint.

4.2.3. Symmetric and antisymmetric operators

Let L be an operator defined in H. If $(L\phi,\psi) = (\phi,L\psi)$ for all ϕ, $\psi\varepsilon$ the domain of L, then L is said to be symmetric. If, on the other hand,

$$(L\phi,\psi) = (\phi,(-L)\psi), \tag{4.12}$$

L is said to be antisymmetric.

Thus for a symmetric operator $L, L^* = L$ and for an antisymmetric operator $L, L^* = -L$. From the above definition it is clear that a self-adjoint operator is necessarily symmetric.

4.2.4. Unitary operator

An operator U defined for every element of H is said to be unitary if

$$UU^* = U^*U = I, \tag{4.13}$$

where I is the identity operator.

Thus for an unitary operator U

$$U^* = U^{-1}, \quad U = (U^*)^{-1} \tag{4.14}$$

$$U = U^{**} = (U^{-1})^*$$

where we have used the general property of the operators, namely

$$T^{**} = T. \tag{4.15}$$

We shall presently use the following property of the operators also. Let T and S be two operators defined over H, then

$$(TS)^* = S^*T^* \tag{4.16}$$

4.2.5. Unitary equivalence

An operator $L(t)$ is said to be 'unitarily equivalent', if there exists a one-parameter family of unitary operators $U(t)$ such that

$$U^{-1}(t)L(t)U(t) \tag{4.17}$$

is independent of t.

Mathematically, we express this fact as follows:

$$\frac{d}{dt}[U^{-1}(t)L(t)U(t)] = 0$$

or

$$[U^{-1}]_t LU + U^{-1}L_t U + U^{-1}LU_t = 0, \qquad (4.18)$$

where suffix t denotes derivation with respect to t.

Example: let $D = \dfrac{d}{dt}$ and $(\phi, \psi) = \displaystyle\int_{-\infty}^{\infty} \phi\psi \, dt$ is the L_2-inner product defined over the Hilbert space H and $\phi, \psi \, \varepsilon \, H$.

Then

$$(D\phi, \psi) = \int_{-\infty}^{\infty} \frac{d\phi}{dt}\psi \, dt$$

$$= -\int_{-\infty}^{\infty} \phi\frac{d\psi}{dt} dt = (\phi, -D\psi).$$

Therefore, D is an antisymmetric operator.

Let us now consider

$$(D^2\phi, \psi) = \int_{-\infty}^{\infty} \frac{d^2\phi}{dt^2}\psi \, dt$$

$$= \int_{-\infty}^{\infty} \phi\frac{d^2\psi}{dt^2} dt = (\phi, D^2\psi).$$

Thus D^2 is a symmetric operator.

In general, for this inner product, we can easily show that the following are respectively the antisymmetric and symmetric operators of order $2q + 1$ and $2q$:

$$B_{2q+1} = D^{2q+1} + \sum_{j=1}^{q} (b_j D^{2j-1} + D^{2j-1} b_j) \qquad (4.19a)$$

and

$$B_{2q} = D^{2q} + \sum_{j=1}^{q} (b_j D^{2j} + D^{2j} b_j) \qquad (4.19b)$$

We *note* that each of these operators contain q unknown functions b_j, $j = 1, 2, \ldots, q$.

We *further note* that we could directly prove that D^2 is symmetric without actually performing the integration:

$$(D^2\phi, \psi) = (D(D\phi), \psi) = (D\phi, (-D)\psi)$$

$$= (\phi, (-D)(-D)\psi)$$

$$= (\phi, D^2\psi). \qquad (4.20)$$

so that $(D^2)^* = D^2$.

In a general case when we do not specify a particular norm, we can still prove the above results if we impose the condition that the inner product, i.e., the bilinear functiona (,) is invariant under translation,

i.e.
$$(f(x), g(x)) = (f(x + z), g(x + z)).$$

Differentiating it with respect to the parameter z and putting $z = 0$, we get

$$(Df, g) + (f, Dg) = 0$$

or

$$(Df(x), g(x)) = (f(x), (-D)g(x))$$

i.e.

$$D^* = -D. \tag{4.21}$$

The results for higher order derivatives can be easily obtained by the method indicated above.

We now prove the following theorem to explain Definition (4.2.5).

Theorem 1. If $U(t)$ is an unitary operator defined over a Hilbert space, then $U_t = BU$, where B is an antisymmetric operator depending on t.

Proof. Since $UU^* = U^*U = I$, we have after differentiating with respect to t,

$$U_t U^* + U(U^*)_t = 0 \text{ and } (U^*)_t U + U^* U_t = 0$$

so that

$$U_t = -U(U^*)_t (U^*)^{-1} = -U(U^*)_t U \equiv BU \tag{4.22}$$

and

$$(U^*)_t = -U^* U_t U^*, \tag{4.23}$$

where

$$B = -U(U^*)_t,$$

so that

$$B^* = -[(U^*)_t]^* U^* = -U_t U^*$$

provided $[(U^*)_t]^* = U_t$, i.e. provided the operations of forming adjoint and t-derivation are invertible.

We prove that this is so.

$$(U\phi, \psi) = (\phi, U^*\psi), \quad \phi, \psi \varepsilon H,$$

therefore by differentiating with respect to t we have

$$(U_t \phi, \psi) = (\phi, (U^*)_t \psi).$$

Also by the definition of adjoint of U_t, we have

$$((U_t)\phi,\psi) = (\phi,(U_t)^*\psi).$$

Therefore $(U^*)_t = (U_t)^*$ i.e., $[(U^*)_t]^* = U_t.$

Now from (4.23), we have

$$B^* = -U_tU^* = (U^*)^{-1}(U^*)_t = U(U^*)_t = -B$$

on using properties of $U(t)$ derived in §4.2.4.

Therefore $B^* = -B.$

This completes the proof of the Theorem. The converse of the theorem is also true.

One important consequence of this result is that the unitary operators form a one-parameter family.

We now establish a theorem which will form the basic argument in our discussion.

Theorem 2. If $L(t)$ is a unitarily equivalent operator, then there esists an antisymmetric operator B depending on t such that

$$L_t = [B,L], \tag{4.24}$$

where $[B,L] = BL - LB$ denotes the commutator of B with L.

Proof. $L(t)$ is a unitarily equivalent operator, so there exists a one-parameter family of operators satisfying (4.18). We first calculate $(U^{-1})_t$.

Let $F = Uf$, so that $f = U^{-1}F$.

Therefore $f_t = [U^{-1}]_t F + U^{-1}F_t$

$$= [U^{-1}]_t F + U^{-1}[U_t f + Uf_t]$$

or $[U^{-1}]_t F = -U^{-1}U_t f = -U^{-1}U_tU^{-1}F.$

Therefore we have

$$[U^{-1}]_t = -U^{-1}U_tU^{-1}. \tag{4.25}$$

Substituting eqns (4.22) and (4.25) in eqn (4.18), we get

$$-U^{-1}BLU + U^{-1}L_tU + U^{-1}LBU = 0$$

or $L_t = BL - LB = [B,L].$

This completes the proof of the theorem.

The concept of *unitary equivalence* is extremely important for our purpose. Let L remain a unitarily equivalent operator as t changes and

$$L\psi = \lambda\psi$$

so that λ is an eigenvalue of the operator L. Then there exists a unitary operator U such that $U^{-1} LU$ is independent of t. Then the eigenvalues of $U^{-1}LU$ are independent of t. Writing $\psi = U\phi$ and pre-multiplying the last equation by U^{-1} we get

$$(U^{-1}LU)\phi = \lambda\phi.$$

From this we conclude that *the eigenvalues of the unitarily equivalent operator L are time-independent.*

Thus we have proved the following basic theorem:

Theorem 3. The eigenvalues of the symmetric operator $\underset{u}{L}$, where u evolves according to the equation $u_t = K(u)$, are independent of t, if $\underset{u}{L}$ is unitarily equivalent. Consequently, eigenvalues λ of $\underset{u}{L}$ are 'integrals' of the equations of evolution.

The crux of the problem is that there exists an antisymmetric operator B which satisfies eqn (4.24). In the case of the Schrödinger equation we shall be able to construct such an antisymmetric operator, while in the general case we shall assume the existence of such an antisymmetric operator.

If the operator $\underset{u}{L}$ can be expressed in the following form:

$$\underset{u}{L} = L_0 + \underset{u}{M},$$

where L_0 is independent of u and $\underset{u}{M}$ depends *linearly on u*, then Theorem 3 can be stated in an alternate form. This sort of situation arises in case of the Schrödinger operator:

$$L = \frac{d^2}{dx^2} + \frac{1}{6}u,$$

which is symmetric and which can be expressed in the above form if we take

$$L_0 \equiv \frac{d^2}{dx^2} \quad \text{(which is independent of } u\text{)}$$

and $\qquad \underset{u}{M} = \frac{1}{6}u$ (which depends linearly on u). $\qquad$ (4.26)

Theorem 4. Let $\underset{u}{L}$ be a symmetric operator depending on u, a solution of

$$u_t = K(u), \qquad (4.27)$$

which is expressible in the following manner:

$$\underset{u}{L} = L_0 + \underset{u}{M}, \qquad (4.28)$$

where L_0 is independent of u and $\underset{u}{M}$ depends linearly on u. Let us assume that *there exists an antisymmetric operator B depending on u such that*

$$[B, \underset{u}{L}] = \underset{K(u)}{M},$$ (4.29)

then the eigenvalues of $\underset{u}{L}$ are 'integrals' of (4.27).

Proof. Dropping the subscript u on $\underset{u}{L}$, we have

$$L\psi = L_0\psi + \underset{u}{M}\psi$$

Differentiating with respect to t, we have

$$L_t\psi + L\psi_t = L_0\psi_t + \underset{u}{M_t}\psi + \underset{u}{M}\psi_t$$

or

$$L_t = \underset{u}{M_t}.$$ (4.30)

Since $\underset{u}{M}$ depends linearly on u, $\underset{u}{M_t}$ depends linearly on u_t i.e. on $K(u)$. Therefore, we have

$$L_t = \underset{K(u)}{M}$$

and then from eqn (4.29), we get

$$L_t = [B, L].$$

Hence L remains unitary equivalent as u evolves according to (4.27). Hence the theorem.

4.3. Solitary-wave solution of the general equation of evolution

We assume the following existence and uniqueness theorems for the equation of evolution

$$u_t = K(u).$$ (4.31)

Existence theorem. Eqn (4.31) admits a solution for a given initial condition $u(x,0)$ provided it is sufficiently smooth and along with its derivatives of all order tends to zero as $|x| \to \infty$.

Uniqueness theorem. Given an initial condition satisfying the condition mentioned in the above theorem, (4.31) admits one and only one solution.

Under the above assumptions, we first associate a linear variational equation with a one-parameter family of solutions of eqn (4.31). We can generate a one-parameter family of solutions of eqn (4.31) by making the initial value a function of a parameter ε:

$$u_\varepsilon(x,0) = u_0(x) + \varepsilon f(x).$$ (4.32)

Corresponding to the initial data (4.32), we have a one-parameter family of solutions $u_\varepsilon(x, t)$ of (4.31) which for small values of ε can be written in the form

$$u_\varepsilon(x, t) = u(x, t) + \varepsilon v(x, t) + O(\varepsilon^2),$$ (4.33)

where $u(x, t)$ is the solution of (4.31) corresponding to the initial value $u_0(x)$ but $v(x, t)$ is not a solution of (4.31). We shall establish an equation governing the evolution of v. This requires the following assumption: the operator $K(u)$ depends differentiably on u, i.e.,

$$\frac{d}{d\varepsilon} K(u + \varepsilon v)|_{\varepsilon = 0} = V(u)v \tag{4.34}$$

exists uniquely and is a linear function of v. $V(u)$ is called the *variation* of K. We note that the left-hand side of (4.34) is the derivative of $K(u)$ in the *Frechet sense*. In the sequel, we shall denote the Frechet derivative of a function by putting a dot above it.

Differentiating

$$(u + \varepsilon v)_t = K(u + \varepsilon v)$$

with respect to ε and setting $\varepsilon = 0$, we have, in view of (4.34),

$$v_t = V(u)v. \tag{4.35}$$

Eqn (4.35) is called the *variational equation* for

$$v = \frac{du_\varepsilon}{d\varepsilon}\bigg|_{\varepsilon = 0} \tag{4.36}$$

Let $I(u)$ be an *integral* of (4.31). We *assume* that $I(u)$ is differentiable in the Frechet sense, i.e.,

$$\frac{d}{d\varepsilon} I(u + \varepsilon v)\bigg|_{\varepsilon = 0} \tag{4.37}$$

exists and is a linear functional of v, which we represent by $(G(u), v)$ so that we can write

$$\dot{I} = (G(u), v). \tag{4.38}$$

$G(u)$ is called the *gradient* of $I(u)$.

Since $I(u)$ is an integral of (4.31), $I(u_\varepsilon(t))$ is independent for t for every value of ε and therefore $(G(u), v)$ is also independent of t. Thus we have proved the following theorem.

Theorem 5. Let $u(t)$ be any solution of (4.31) and $v(t)$ be any solution of the corresponding variational equation. Let $I(u)$ be an integral of (4.31) and $G(u)$ its gradient, then the bilinear functional $(G(u), v)$ is independent of t.

N.B. The above result corresponds to the following result for a linear equation. Let us consider the linearized version

$$u_t + cu_x + u_{xxx} = 0$$

of the KdV equation. Multiplying by u and integrating with respect to x from

$x = -\infty$ to $x = \infty$, we get

$$\frac{d}{dt} \int_{-\infty}^{\infty} \frac{1}{2} u^2 \, dx = 0$$

assuming that u, $u_{xx} \to 0$ as $|x| \to \infty$. Thus, the quadratic functional $Q(u,u)$ $= \int_{-\infty}^{\infty} \frac{1}{2} u^2 \, dx$ is independent of t. We shall call it *energy*. Let now u and v be any two solutions of the above equation. Then by the *superposition principle* $u \pm v$ are also the solutions of the equation.

Hence the quadratic functionals $Q(u+v, u+v)$ and $Q(u-v, u-v)$ are independent of t. Therefore, on subtraction, we conclude that

$$\int_{-\infty}^{\infty} \left\{ \frac{1}{2}(u+v)^2 - \frac{1}{2}(u-v)^2 \right\} dx = 2 \int_{-\infty}^{\infty} uv \, dx$$
$$= 2(u,v),$$

which is bilinear functional of u and v, is also independent of t.

We shall now prove the following basic theorem of Lax:

Theorem 6. *Let us assume that the equation of evolution*

$$u_t = K(u)$$

satisfies the following conditions:

(a) *$K(u)$ depends differentiably on u, its variation is $V(u)$;*

(b) *the equation is invariant under translation and preserves a positive translation-invariant quadratic functional, which we call energy;*

(c) *the equation has a solitary-wave solution;*

(d) *the only function annihilated by $cD - V^*(s)$ and vanishing at $\pm \infty$ is a multiple of s.*

Let $I(u)$ be an integral of the equation such that

(e) *it is differentiable in the Frechet sense, and*

(f) *its gradient $G(u)$ vanishes at $\pm \infty$ for $u = s$.*

Then $G(s) = \beta s,$

where β depends on I and c, i.e., every solitary wave is an eigenfunction of the gradient of an integral of the equation of evolution.

Proof. Let v be any solution of the variational equation (Assumption (a)) and $s(\xi)$, $\xi = x - ct$, be a solitary wave solution of (4.31) (Assumption (c)). Then by Theorem 5

$$(G(s(x - ct), v(x,t)) \tag{4.39}$$

is independent of t.

In view of translation invariance of (,) (Assumption (b)), we have

$$(G(s(x)), \ v(x + ct, t)) \tag{4.40}$$

is independent of t.

Let us write for brevity

$$v(x + ct, t) = w(x, t), \tag{4.41}$$

then (4.40) may be written as

$$(G(s(x)), \ w(x, t)) \tag{4.42}$$

which is independent of t.

Differentiating (4.42) with respect to t and noting that x and t are independent variables, we have

$$(G(s(x)), w_t) = 0, \tag{4.43}$$

where from (4.41)

$$w_t = v_t + c v_X,$$

and

$$w_x = v_X, \qquad \text{where} \qquad X = x + ct$$

so that

$$w_t = v_t + c w_x$$
$$= V(s(x))w + c w_x$$
$$= \{cD + V(s(x))\}w, \quad D \equiv \frac{\partial}{\partial x}. \tag{4.44}$$

Substituting (4.44) in (4.43), we have

$$(G(s(x)), \{cD + V(s(x))\}w) = 0. \tag{4.45}$$

Writing it in the adjoint form, we have

$$(\{-cD + V^*(s(x))\}G(s), w) = 0, \tag{4.46}$$

where V^* is the operator adjoint to V.

The value of w at any particular time, say $t = 0$, can be prescribed arbitrarily, therefore, from (4.46), we have

$$\{-cD + V^*(s(x))\}G(s) = 0. \tag{4.47}$$

By assumption (b), (4.31) is energy preserving, i.e., $(u(t), u(t))$ is independent of t so that differentiating with respect to t we have

$$2(u(t), u_t(t)) = 0$$

or

$$2(u, K(u)) = 0.$$

The initial value of u is arbitrary. We substitute u_ε for u. Differentiating with respect to ε and putting $\varepsilon = 0$, we have

$$(v, K(u)) + (u, V(u)v) = 0,$$

or

$$(v, K(u)) + (V^*(u)u, v) = 0,$$

or

$$(v, K(u)) + (v, V^*(u)u) = 0,$$

or

$$(v, K(u) + V^*(u)u) = 0.$$

Since v is *any solution* of the variational equation *we may treat it as arbitrary* at a time t, therefore, the above equation implies that

$$K(u) + V^*(u)u = 0. \tag{4.48}$$

Since we are considering a solitary-wave solution, we write $s(\xi)$, $\xi = x - ct$ for $u(x, t)$ and then from (4.31), we have

$$cs_\xi + K(s(\xi)) = 0. \tag{4.49}$$

Substituting $s(\xi)$ for u and for $K(s(\xi))$ from (4.49) in (4.48), we have

$$- cs_\xi + V^*(s(\xi))s(\xi) = 0$$

or

$$\left[-c\frac{d}{d\xi} + V^*(s(\xi)) \right] s(\xi) = 0. \tag{4.50}$$

Therefore *the solitary wave belongs to the null space of the linear operator*

$$c\frac{d}{d\xi} - V^*(s(\xi)). \tag{4.51}$$

Comparing eqns (4.47) and (4.50) we have, by assumptions (d) and (f),

$$G(s(\xi)) = \beta s(\xi), \tag{4.52}$$

where β depends on c and I as G is the gradient of the integral $I(s(\xi))$.

N.B. The boundary conditions of $G(s)$ and s do not create any difficulty as by assumptions (d) and (f) both tend to zero as $|\xi| \to \infty$.

This completes the proof of Theorem 6.

4.4. Application of the general theory to the KdV equation

We shall apply the above theory to the KdV equation to examine if we obtain the results already discussed.

$$u_t + uu_x + u_{xxx} = 0$$

Therefore,

$$K(u) = -(uu_x + u_{xxx}). \tag{4.53}$$

We denote the one-parameter family of solutions of this equation by

$$u_\varepsilon(x, t) = u(x, t, \varepsilon)$$

which is uniquely determined by the one-parameter family of initial conditions

$$u_\varepsilon(x, 0) = u_0 + \varepsilon f.$$

The variation of $K(u)$ is given by

$$
\begin{aligned}
\frac{\mathrm{d}}{\mathrm{d}\varepsilon} K(u + \varepsilon v)\big|_{\varepsilon = 0} &= -\frac{\mathrm{d}}{\mathrm{d}\varepsilon} \{(u + \varepsilon v)(u_x + \varepsilon v_x) + u_{xxx} + \varepsilon v_{xxx}\}\big|_{\varepsilon = 0} \\
&= -(uv_x + u_x v + v_{xxx}) \\
&\equiv V(u)v, \text{ in our notation}
\end{aligned}
$$

so that

$$V(u) \equiv -(u_x + uD + D^3), \quad D = \frac{\partial}{\partial x} \tag{4.54}$$

$$= -(Du + D^3).$$

From this we conclude that $K(u)$ depends differentiably on u. Therefore $K(u)$ *satisfies condition (a) of Theorem 6.*

For further use, we shall evaluate adjoint V^* of V. Before we can do so we shall have to fix the definition of inner product. Since we are dealing with C^∞ class of functions over $(-\infty < x < \infty)$ it is sufficient if we take the L_2 scalar product. Thus we shall take

$$(u, v) \equiv \int_{-\infty}^{\infty} uv\,\mathrm{d}x. \tag{4.55}$$

From the above definition of the inner product, we have

$$
\begin{aligned}
(V(u)f, g) &= -\int_{-\infty}^{\infty} \left(\frac{\mathrm{d}}{\mathrm{d}x}(uf) + \frac{\mathrm{d}^3 f}{\mathrm{d}x^3} \right) g\,\mathrm{d}x \\
&= \int_{-\infty}^{\infty} f\left(u\frac{\mathrm{d}g}{\mathrm{d}x} + \frac{\mathrm{d}^3 g}{\mathrm{d}x^3} \right)\mathrm{d}x
\end{aligned}
$$

since f, g, f_x, g_x, f_{xx}, $g_{xx} \to 0$ as $|x| \to \infty$. Therefore, we have

$$V^*(u) = uD + D^3. \tag{4.56}$$

From (4.54), the variational equation determining v here is

$$v_t = -(Du + D^3)v.$$

Let us now identify L with the Schrödinger operator:

$$L \equiv \frac{\mathrm{d}^2}{\mathrm{d}x^2} + \frac{1}{6}u, \tag{4.57}$$

then L is clearly symmetric.

Therefore, by Theorem 3, the eigenvalues of L, where u evolves according to the KdV equation, are independent of t, provided there exists an antisymmetric operator B such that

$$L_t = [B, L]. \tag{4.58}$$

Now, on differentiating L with respect to t, we get

$$L_t = \frac{\partial}{\partial t}(D^2 + \tfrac{1}{6}u) = \tfrac{1}{6}u_t = \tfrac{1}{6}K(u)$$

$$= -\tfrac{1}{6}(uu_x + u_{xxx}). \tag{4.59}$$

From eqns (4.58) and (4.59), it is clear that the Theorem 3 will be valid in the present case if

$$[B, D^2 + \tfrac{1}{6}u] = -\tfrac{1}{6}(uu_x + u_{xxx}). \tag{4.60}$$

Since on the right-hand side of (4.60) a third order derivative occurs, we assume the following antisymmetric third-order operator for B

$$B = D^3 + bD + Db.$$

By performing actual differentiation, we can show that

$$[D^3 + bD + Db, D^2 + \tfrac{1}{6}u]\psi$$
$$= (D^3 + bD + Db)(\psi_{xx} + \tfrac{1}{6}u\psi) - (D^2 + \tfrac{1}{6}u)(\psi_{xxx} + b\psi_x + (b\psi)_x)$$
$$= (\tfrac{1}{2}u_x - 4b_x)\psi_{xx} + (\tfrac{1}{2}u_{xx} - 4b_{xx})\psi_x$$
$$+ (\tfrac{1}{6}u_{xxx} + \tfrac{1}{3}bu_x - b_{xxx})\psi. \tag{4.61}$$

In order that it is simply a scalar multiple of the right-hand side of (4.60), we must choose

$$b = \tfrac{1}{8}u \tag{4.62}$$

and then

$$[B, D^2 + \tfrac{1}{6}u] = \tfrac{1}{24}(uu_x + u_{xxx}), \tag{4.63}$$

so that (4.60) will be established if we would have taken

$$B = 4(D^3 + bD + Db).$$

This proves the existence of the antisymmetric operator B satisfying (4.58). Therefore Theorem 3 is valid here and the eigenvalues λ of the Schrödinger operator (4.57) are independent of t.

We note the striking elegance of the above proof. The proof of the same result when it was first discovered was heavily based on physical arguments.

We shall now find the gradient $G(u(t))$ of the integral $\lambda(u(t))$.

According to the definition (4.38), we have

$$\frac{d}{d\varepsilon}\lambda(u+\varepsilon v)|_{\varepsilon=0}=(G(u),v) \tag{4.64}$$

where λ is determined by $L\psi=\lambda\psi$.
Taking the Frechet derivative of the last equation, we have

$$\dot{L}\psi+L\dot{\psi}=\dot{\lambda}\psi+\lambda\dot{\psi} \tag{4.65}$$

where

$$\dot{L}=\frac{d}{d\varepsilon}\left\{\frac{d^2}{dx^2}+\frac{1}{6}(u+\varepsilon v)\right\}\bigg|_{\varepsilon=0}=\frac{1}{6}v. \tag{4.66}$$

Therefore eqn (4.65) reduces to

$$\tfrac{1}{6}v\psi+L\dot{\psi}=\dot{\lambda}\psi+\lambda\dot{\psi}.$$

Multiplying this equation by ψ and integrating with respect to x from $x=-\infty$ to ∞, we have

$$\dot{\lambda}=\int_{-\infty}^{\infty}\frac{1}{6}v\psi^2\,dx=\left(\frac{1}{6}\psi^2,v\right) \tag{4.67}$$

on using the following

$$\int_{-\infty}^{\infty}\psi^2\,dx=1 \quad \text{(normality condition for } \psi\text{)}$$

and

$$\int_{-\infty}^{\infty}L[\psi]\psi\,dx=\int_{-\infty}^{\infty}\psi L[\psi]\,dx \qquad \text{(L is symmetric).}$$

Therefore, from (4.64) and (4.67), we have

$$G(u)=\tfrac{1}{6}\psi^2. \tag{4.68}$$

Thus $\lambda(u)$ satisfies the conditions (e) and (f) of Theorem 6, namely, $\lambda(u)$ is differentiable in Frechet sense and $G(s)\to0$ as $|x|\to\infty$ since $\psi\to0$ as $|x|\to\infty$.

The condition (c) of Theorem 6 also holds because we have shown in Chapter 3 the existence of the solitary-wave solution of the KdV equation.

The condition (b) of Theorem 6 also holds for the KdV equation as the quadratic functional (u,u) is independent of t.

We now prove that the condition (d) of Theorem 6 also holds in the present case. Let X be a function that is annihilated by $cD-V^*(s)$ such that $X\to0$ as $\xi=x-ct\to\pm\infty$. Then we have

$$c\frac{dX}{d\xi}-s(\xi)\frac{dX}{d\xi}-\frac{d^3X}{d\xi^3}=0. \tag{4.69}$$

Now, if we transform the KdV equation to ξ coordinate, we have, for a solitary-wave solution $u(x,t) = s(\xi)$,

$$c\frac{ds}{d\xi} - s(\xi)\frac{ds}{d\xi} - \frac{d^3s}{d\xi^3} = 0, \tag{4.70}$$

where $s \to 0$ as $\xi \to \pm\infty$.

Therefore, from eqns (4.69) and (4.70), we have

$$X = \alpha s, \tag{4.71}$$

where α is a constant.

The foregoing discussion shows that the KdV equation as well as the integral $\lambda(u)$ satisfy all the conditions of Theorem 6. Therefore we have

$$G(s) \equiv \tfrac{1}{6}\psi^2 = \beta s, \tag{4.72}$$

where β depends on $\lambda(s)$ and c.

From (4.72), we have

$$\psi = \sqrt{(6\beta)}s^{1/2}. \tag{4.73}$$

To establish the relation between the integral $\lambda(s)$ which is an eigenvalue of the Schrödinger operator $L^2 + \tfrac{1}{6}s$, where s is the solitary-wave solution of the KdV equation and the eigenspeed $c(s)$, we proceed as follows. Twice differentiating the eigenfunction ψ (of the Schrödinger operator) given by (4.73), we have

$$\psi_x = \tfrac{1}{2}\sqrt{(6\beta)}s^{-1/2}s_\xi$$

and

$$\psi_{xx} = \tfrac{1}{2}\sqrt{(6\beta)}\left[s^{-1/2}s_{\xi\xi} - \tfrac{1}{2}s^{-3/2}s_\xi^2\right]$$

so that

$$\psi_{xx} + \tfrac{1}{6}s\psi = \sqrt{(6\beta)}\tfrac{1}{4}c(s)s^{1/2}, \tag{4.74}$$

on substituting the following values of $s_{\xi\xi}$ and s_ξ^2 obtained by successively integrating (4.70) twice and using the boundary conditions on s and its derivatives at $\xi = \pm\infty$:

$$s_{\xi\xi} = cs - \tfrac{1}{2}s^2$$

and

$$s_\xi^2 = cs^2 - \tfrac{1}{3}s^3.$$

The left-hand side of (4.74) is also equal to $\lambda(s)\psi$

$$\lambda(s)\psi = \sqrt{(6\beta)}\lambda(s)s^{1/2}. \tag{4.75}$$

Therefore, from eqns (4.74) and (4.75), we have the important result:

$$c(s) = 4\lambda(s) = 4\kappa^2(s) \tag{4.76}$$

which we have proved in Chapter 3 using the 'inverse scattering method'. Actually in chapter 3 we proved a more general result: if the solution $u(x,t)$ of the KdV equation is a reflectionless potential of the Schrödinger operator, then corresponding to the eigenvalue κ_p^2 of the Schrödinger operator $\dfrac{d^2}{dx^2}$ $+\frac{1}{2}u$, there corresponds a soliton moving with speed $4\kappa_p^2$ at infinity. We have yet to relate these eigenspeeds with an arbitrary solution $u(x,t)$ of the KdV equation which need not be reflectionless but which has these individual solitons embedded in it and which show up only in the asymptotic behaviour of $u(x,t)$. We shall take up converse of this problem in the next section.

4.5. Eigenspeeds of the general solution of the KdV equation

We shall now prove that when c is an eigenspeed of a solution $u(x,t)$, reflectionless or not, of the KdV equation $\frac{1}{4}c$ is an eigenvalue of the Schrödinger operator $\dfrac{d^2}{dx^2}+\dfrac{1}{6}u$.

Let $u(x,t)$ be any solution of the KdV equation which evolves asymptotically into a number of solitary waves. Let us consider one of these solitary waves $u = s(\xi)$, $\xi = x - ct$, moving with speed c, i.e.,

$$\lim_{\substack{t \to \pm\infty \\ \xi \text{ fixed}}} u(x,t) = s(\xi - \theta^\pm), \tag{4.77}$$

where $\theta^\pm$ are constants. Therefore, given arbitrary positive numbers ε and X, there exists a positive number $T(X,\varepsilon)$ such that

$$|u(\xi + ct, t) - s(\xi - \theta)| \le \varepsilon \tag{4.78}$$

for all $t : |t| > T$ and for all $\xi : |\xi| < X$. $\tag{4.79}$

We note that X can be as large as we please. However, $\lim_{X \to \infty} T(X,\varepsilon) = \infty$ for fixed ε.

Writing $t = T$, (4.78) reduces to

$$|u(\xi + cT, T) - s(\xi - \theta)| < \varepsilon \Big\}$$
for all $\qquad \xi : |\xi| < X$ or $cT - X < x < cT + X.\Big\} \tag{4.80}$

Let L_T denote the Schrödinger operator $\dfrac{d^2}{dx^2}+\dfrac{1}{6}u(x, T)$ written at $t = T$. We shall first prove that $\frac{1}{4}c$ is an approximate eigenvalue and

$$\psi_T = s^{1/2}(x - cT - \theta)$$

the corresponding approximate eigenfunction of L_T in the sense that

$$\|L_T\psi_T - \tfrac{1}{4}c\psi_T\| < \delta\|\psi_T\|, \tag{4.81}$$

where $||\ldots||$ denotes the L_2 norm and $\delta \to 0$ as $\varepsilon \to 0$ and $X \to \infty$. We note that ψ_T is the eigenvector, which corresponds to the eigenvalue $\frac{1}{4}c$, of $\dfrac{d^2}{dx^2} + \dfrac{1}{6}s(x - cT - \theta)$ and not of L_T. In general, ψ_T is not an eigenvector of L_T.

Now,

$$
\begin{aligned}
|L_T\psi_T - \tfrac{1}{4}c\psi_T| &= \left| \left\{ \frac{d^2}{dx^2} + \frac{1}{6}u(x, T) - \frac{1}{4}c \right\} \psi_T \right| \\
&= \left| \left\{ \frac{d^2}{dx^2} + \frac{1}{6}s(x - cT - \theta) - \frac{1}{4}c \right\} \psi_T \right. \\
&\quad \left. + \frac{1}{6} \left\{ u(x, T) - s(x - cT - \theta) \right\} \psi_T \right| \\
&= \tfrac{1}{6}|u(x, T) - s(x - cT - \theta)|\psi_T < \tfrac{1}{6}\varepsilon\psi_T
\end{aligned}
\qquad (4.82)
$$

from (4.80) provided $cT - X \leq x \leq cT + X$. In this interval eqn (4.82) provides an estimate for $L_T\psi_T - \frac{1}{4}c\psi_T$. We shall now obtain an estimate for this expression outside this interval.

In the range outside the interval $(cT - X, cT + X)$, we proceed as follows. Assuming that the solutions of the KdV equation are uniformly bounded with respect to time, we denote by M the supremum of $u(x, t)$. Then from

$$
|L_T\psi_T - \tfrac{1}{4}c\psi_T| = \tfrac{1}{6}|u(x, T) - s(x - cT - \theta)|\psi_T
$$

and the expression (4.8) for s we get

$$
|L_T\psi_T - \tfrac{1}{4}c\psi_T| < \tfrac{1}{6}(M + 3c)\psi_T.
\qquad (4.83)
$$

In the range $x < cT - X$, using $\psi_T = s^{1/2}$,

$$
\begin{aligned}
\psi_T &= \sqrt{(3c)} \operatorname{sech} \left\{ \frac{\sqrt{c}}{2}(x - cT - \theta) \right\} \\
&= \frac{2\sqrt{(3c)}}{\exp\left\{ \dfrac{\sqrt{c}}{2}(x - cT - \theta) \right\} + \exp\left\{ -\dfrac{\sqrt{c}}{2}(x - cT - \theta) \right\}} \\
&< 2\sqrt{3c}\, \exp\left\{ \frac{\sqrt{c}}{2}(x - cT - \theta) \right\}.
\end{aligned}
\qquad (4.84)
$$

Similarly for $x > cT + X$, we get

$$
\psi_T < 2\sqrt{(3c)} \exp\left\{ -\frac{\sqrt{c}}{2}(x - cT - \theta) \right\}.
\qquad (4.85)
$$

Therefore, according to the choice of our norm

$$||L_T\psi_T - \tfrac{1}{4}c\psi_T||^2 = \left[\int_{-\infty}^{cT-X} + \int_{cT-X}^{cT+X} + \int_{cT+X}^{\infty}\right]\{L_T\psi_T - \tfrac{1}{4}c\psi_T\}^2 dx.$$

From eqns (4.82), (4.83), (4.84) and (4.85), we get

$$||L_T\psi_T - \tfrac{1}{4}c\psi_T||^2 \le \tfrac{1}{36}\varepsilon^2 \int_{cT-X}^{cT+X} \psi_T^2 dx$$

$$+ \tfrac{1}{3}c(M + 3c)^2\left[\int_{-\infty}^{cT-X} \exp\left\{\sqrt{c}\,(x - cT - \theta)\right\}dx\right.$$

$$+ \left.\int_{cT+X}^{\infty} \exp\left\{-\sqrt{c}\,(x - cT - \theta)\right\}dx\right],$$

or

$$||L_T\psi_T - \tfrac{1}{4}c\psi_T||^2 < (\tfrac{1}{36}\varepsilon^2 + \tfrac{1}{3}(\sqrt{c})(M + 3c)^2[\exp\{-\sqrt{c}(X + \theta)\}$$
$$+ \exp\{-\sqrt{c}(X - \theta)\}]/||\psi_T||^2)\cdot||\psi_T||^2.$$

Since $||\psi_T||$ is non zero and finite, the above result implies inequality (4.81) where $\delta \to 0$ as $\varepsilon \to 0$ and $X \to \infty$.

Thus $\dfrac{c}{4}$ lies within δ of a point of the spectrum of the operator $L_T \equiv \dfrac{d^2}{dx^2}$

$+ \dfrac{1}{6}u(x, T)$. Since $\delta \to 0$ as $\varepsilon \to 0$ and $X \to \infty$ and eigenvalue of L_T is

independent of T, we conclude that $\dfrac{c}{4}$ is an eigenvalue of the opera-

tor $L \equiv \dfrac{d^2}{dx^2} + \dfrac{1}{6}u(x, t)$.

Bibliography

LAX, P. D. Integrals of nonlinear equations of evolution and solitary waves. *Communs pure appl. Math.* **21**, 467–90 (1968).

LUSTERNIK, L. A. and SOBOLEV, V. I. *Elements of functional analysis.* Hindustan Publishing Corporation, Delhi (1974).

MIURA, R. M. GARDNER, C. S. and KRUSKAL, M. D. Korteweg–de Vries equation and generalizations. II. Existence of conservation laws and constants of motion. *J. Math. Phys.* **9**, 1204–9 (1968).

PRICE, J. D. *Basic methods of linear functional analysis.* Hiltchinson University Library, London (1973).

SJOBERG, A. On the Korteweg–de Vries equation. Uppasala University Department of Computer Science Report (1967).

GROUP VELOCITY : NONLINEAR WAVES

5.1. Introduction

Our discussion of the linear waves in Chapter 1 has brought out the following facts:

(i) a linear wave in a homogeneous conservative dispersive medium which is initially uniform (i.e., with k, ω, a, V_p and V_g independent of x and t initially) after a sufficiently long time $(t \gg P)$ develops into a non-uniform wave-train of increasing length along which k, ω, a, V_p, V_g vary slowly with x and t (in fact in the combination x/t);

(ii) the significant changes (i.e., changes of $O(1)$) in these parameters take place over the length and time scales of the order of $L = O(x)$ and $T = O(t)$ respectively;

(iii) we could, therefore, still treat the wave harmonic with slowly varying parameters over the length and time scales of the order of X and $\tau : \lambda \ll X \ll L, P \ll \tau \ll T$, where λ is the wavelength and P the period of the original uniform wave;

(iv) we have developed the mathematical expression, $V_g = \omega'(k)$, for the group velocity through the dispersion relation subjecting the governing equation to the Fourier analysis;

(v) the wave number, frequency and the energy of the wave propagate with the group velocity, a concept to which we have given *physical* meaning only through the asymptotic behaviour of the exact solution of the initial value problem as $t \to \infty$.

The nonlinear-wave equation cannot be resolved into its Fourier components; we have to evolve some method of defining even the parameters like wave number, period, etc. The nonlinear equations do not admit harmonic solutions of the type

$$u(x,t) = \int_0^\infty A(k) \exp[i\{kx - \omega(k)t\}]dk,$$

however, in Chapter 2, we have seen that they do admit elementary solutions which may represent the uniform wave-trains in the moving coordinate $\xi = x - ct$, where c is the velocity of the moving coordinate. It is through the study of these *steady* solutions that we shall be able to give meanings to the parameters: wave number, frequency, group velocity, etc.

5.2. Averaging procedure

Our study of the KdV equation has shown that it admits steady solutions of the type

which gives

$$\left.\begin{array}{c} \Phi_\xi^2 = F(\Phi;c,A_i) \\[2mm] \Phi = \Phi(\xi;c,A_i), \quad \xi = x - ct \end{array}\right\} \tag{5.1}$$

where A_i are constants of integration whose number depend on the order of the equation with respect to the space derivative. In the present chapter, we shall discuss nonlinear equations of only those type which admit steady solutions of the type (5.1).

It is clear that the solution is oscillatory between two consecutive zeros, say $\Phi_1(c,A_i)$ and $\Phi_2(c,A_i), (\Phi_2 > \Phi_1)$ of the function F between which it is positive definite. The condition of positive definiteness between Φ_1 and Φ_2 is necessary to ensure that Φ_ξ is real.

Let ξ_1 and ξ_2 be the values of ξ at which

and

$$\left.\begin{array}{c} \Phi(\xi_1; c, A_i) = \Phi_1(c, A_i) \\[2mm] \Phi(\xi_2; c, A_i) = \Phi_2(c, A_i) \end{array}\right\} \tag{5.2}$$

Then, from the analogy of the linear wave, we may define the wavelength λ for a nonlinear wave as

$$\lambda = \lambda(c, A_i) = 2 \int_{\xi_1}^{\xi_2} d\xi \tag{5.3a}$$

$$= 2 \int_{\Phi_1}^{\Phi_2} \frac{d\Phi}{\Phi_\xi} \tag{5.3b}$$

$$= 2 \int_{\Phi_1}^{\Phi_2} \frac{d\Phi}{F^{1/2}(\Phi;c,A_i)} \tag{5.3c}$$

and then, in the usual manner, the wave number k is defined as

$$k = k(c, A_i) = \frac{2\pi}{\lambda(c, A_i)} \tag{5.4}$$

and frequency ω as

$$\omega = \omega(c, A_i) = ck(c, A_i). \tag{5.5}$$

We have the following expression for the period of the wave:

$$P = P(c, A_i) = \frac{2\pi}{\omega(c, A_i)}. \tag{5.6}$$

For sake of brevity of writing we shall not always explicitly show the dependence of k, ω, λ, P on c and A_i.

The solution (5.1), where c and A_i are constants, describes the wave *locally*. In direct analogy with the linear waves, we can obtain a more general solution by regarding c and A_i as slowly varying functions of x and t. We shall now develop a method for determining this dependence of these parameters on x and t.

First of all we must introduce some averaging procedure to eliminate the rapid fluctuations in the field quantities taking place at the smaller scale ($x \sim \lambda$ and $t \sim P$). Only then we hope to be able to study the slow variations in the parameters c and A_i and through them the variations in k, ω, etc. In order to avoid the loss of significant variations in k, ω, etc. in the averaging procedure, we shall restrict our time and space intervals which are much smaller than T and L. The foregoing discussion suggests the function of an intermediary scale, X and τ, so that

$$\lambda \ll X \ll L, \quad P \ll \tau \ll T \tag{5.7}$$

and then we define the average $\tilde{F}(x, t)$ at a general point x for a fixed value of t by the relation

$$\tilde{F}(x, t) = \frac{1}{2X} \int_{x-X}^{x+X} F(x', t) \, dx'. \tag{5.8}$$

Whitham chose the conservation forms of the governing equations for averaging purpose. Therefore, we take the conservation law

$$P_t + Q_x = 0. \tag{5.9}$$

We subject this to averaging procedure to obtain differential equations for determining c and A_i which do not contain x and t explicitly.

Now we have

$$\tilde{P}_t = \frac{1}{2X} \int_{x-X}^{x+X} \frac{\partial}{\partial t} P(x', t) \, dx' = (\tilde{P})_t \tag{5.10a}$$

and

$$\tilde{Q}_x = \frac{1}{2X} \int_{x-X}^{x+X} \frac{\partial}{\partial x'} Q(x', t) \, dx'$$

$$= \frac{1}{2X} [Q(x + X, t) - Q(x - X, t)]$$

$$= \frac{\partial}{\partial x} \frac{1}{2X} \int_{x-X}^{x+X} Q(x',t)dx'$$

$$= (\tilde{Q})_x \tag{5.10b}$$

so that on averaging (5.9), we get

$$\frac{\partial}{\partial t} \tilde{P}(x,t;c,A_i) + \frac{\partial}{\partial x} \tilde{Q}(x,t;c,A_i) = 0. \tag{5.11}$$

Whitham points out that the advantage of starting from the conservation law is that both terms in eqn (5.11) are of the same order (λ/L). If an undifferentiated term, say R, were also present, a detailed solution more accurate than (5.1) would be needed to calculate R to order λ/L.

We note that, in eqn (5.11), $\tilde{P}$ and $\tilde{Q}$ still depend explicitly on x and t. To remove this explicit dependence we argue in the following manner. The interval $(x - X, x + X)$, X being much larger than λ, contains a large number of waves over which we may sensibly treat k, ω, etc. as more or less constant in view of (5.7). In this situation, we can replace $\tilde{P}$ and $\tilde{Q}$ respectively by $\bar{P}(c, A_i)$ and $\bar{Q}(c, A_i)$ which are averages obtained by keeping k, ω, etc. constant in $(x - X, x + X)$. It is clear that in making this approximation, we shall be introducing errors of the order of λ/X and X/L, which are small due to the assumption (5.7).

Now we note that the average $\bar{F}$ is approximately the same as the average of F over a period. This is again because there is a large number of uniform waves in $(x - X, x + X)$. Thus, we have

$$\bar{P}(c, A_i) = \frac{1}{\lambda} \int_x^{x+\lambda} P(x',t)dx'$$

$$= \frac{1}{\lambda} \int_0^{\lambda} P(X,t)dX, \quad x' = x + X$$

$$= \frac{1}{\lambda} \int_0^{\lambda} P(\Phi; c, A_i)dX, \tag{5.12}$$

where $P(\Phi)$ indicates functional dependence on Φ.

In terms of averages defined in (5.12), the averaged conservation equation reduces to

$$\frac{\partial}{\partial t} \bar{P}(c, A_i) + \frac{\partial}{\partial x} \bar{Q}(c, A_i) = 0, \tag{5.13}$$

where c and A_i are slowly varying functions of x and t.

Now we know that for certain problems we have an infinite set of conservation equations and hence an infinite set of averaged conservation equations of the type (5.13). Therefore, for the success of the method it is crucial that of this infinite set of averaged conservation equations only as many are independent as the number of the parameters c and A_i which we have to determine. In all the problems that have been studied so far, the number of independent averaged conserved equations is exactly the same as the number of the unknown parameters. However, so far, there is no general proof for this statement and it is surmised that the secret of the proof lies in the transformation properties of the governing system of equations.

Again experience has shown that if, in direct analogy with the adiabatic invariant $I = \oint p\,dq$ in the Hamiltonian mechanics, we introduce a function

$$W(c, A_i) \equiv \oint \Phi_\xi\, d\Phi$$

we are able to express wave number, frequency, and other parameters associated with the wave in terms of the first order partial derivatives of W with respect to c and A_i. Moreover, in many cases the system of partial differential equations determining these partial derivatives of W turn out to be hyperbolic. Consequently, by finding out the characteristics of this hyperbolic system we are able to define the characteristic speeds associated with the wave. The corresponding compatibility relations give the quantities which remain constant along these characteristics. Generally, we have more than one characteristic speed; we shall have to specify which one of them should be designated as the group velocity. The direct analogy with the linear case does not help as all these characteristic speeds become equal to the group velocity. We thus discuss this point again in the context of the examples that we consider in the next section.

5.3. Examples

In this section we shall illustrate the above method given by Whitham by solving some examples which have also been solved by Whitham. We first solve a linear equation to check that Whitham's method is reliable and then solve two nonlinear equations by this method one of them being the KdV equation.

5.3.1. *Linear wave*

We shall first apply the method of §5.2 to the following linear equation:

$$\phi_{tt} - \phi_{xx} + \phi = 0 \tag{5.14}$$

which remains invariant under the transformation $x \to -x$ and $t \to -t$.

To get the steady solution, we set

$$\xi = x - ct \tag{5.15}$$

in (5.14) which reduces to

$$(c^2 - 1)\Phi_{\xi\xi} + \Phi = 0. \tag{5.16}$$

On integration, we have

$$\Phi_\xi = \frac{1}{\sqrt{(c^2 - 1)}} (2A - \Phi^2)^{1/2},$$

following the positive branch and

$$\xi = \sqrt{(c^2 - 1)} \int \frac{d\Phi}{\sqrt{(2A - \Phi^2)}} \tag{5.17}$$

so that

$$\Phi(\xi; c, A) = \sqrt{(2A)} \cos\left(\frac{\xi - \xi_0}{\sqrt{(c^2 - 1)}}\right). \tag{5.18}$$

$\Phi(\xi)$ oscillates between $-\sqrt{(2A)}$ and $\sqrt{(2A)}$ so that we define

$$\lambda(c, A) = 2\sqrt{(c^2 - 1)} \int_{-\sqrt{(2A)}}^{\sqrt{(2A)}} \frac{d\Phi}{\sqrt{(2A - \Phi^2)}}$$

$$= 2\pi\sqrt{(c^2 - 1)} \tag{5.19}$$

and then

$$k(c, A) = \frac{1}{\sqrt{(c^2 - 1)}} \tag{5.20}$$

and

$$\omega(c, A) = \frac{c}{\sqrt{(c^2 - 1)}} = \sqrt{(1 + k^2)}. \tag{5.21}$$

On substitution $\phi \propto \exp\{i(kx - \omega t)\}$ in (5.14), we get the dispersion relation

$$\omega = \pm\sqrt{(1 + k^2)}$$

one of which is the same as (5.21).

On multiplying (5.14) by ϕ_t and ϕ_x and rearranging the terms, we have the following two conservation equations which are sufficient to determine the variations in c and A with respect to x and t:

$$[\tfrac{1}{2}\phi_t^2 + \tfrac{1}{2}\phi_x^2 + \tfrac{1}{2}\phi^2]_t + [-\phi_t\phi_x]_x = 0 \tag{5.22a}$$

and

$$[-\phi_x\phi_t]_t + [\tfrac{1}{2}\phi_t^2 + \tfrac{1}{2}\phi_x^2 - \tfrac{1}{2}\phi^2]_x = 0 \tag{5.22b}$$

which reduce to

$$[\tfrac{1}{2}(c^2 + 1)\Phi_\xi^2 + \tfrac{1}{2}\Phi^2]_t + [c\Phi_\xi^2]_x = 0 \tag{5.22c}$$

and

$$[c\Phi_\xi^2]_t + [\tfrac{1}{2}(c^2+1)\Phi_\xi^2 - \tfrac{1}{2}\Phi^2]_x = 0 \tag{5.22d}$$

in view of (5.15).

Substituting for Φ and Φ_ξ from eqn (5.18) in eqns (5.22 c and d) we get

$$\left\{ \frac{c^2+1}{c^2-1} A - \frac{2A}{c^2-1} \cos^2\left(\frac{\xi-\xi_0}{\sqrt{(c^2-1)}}\right) \right\}_t$$

$$+ \left\{ \frac{2Ac}{c^2-1} \sin^2\left(\frac{\xi-\xi_0}{\sqrt{(c^2-1)}}\right) \right\}_x = 0 \tag{5.22e}$$

and

$$\left\{ \frac{2Ac}{c^2-1} \sin^2\left(\frac{\xi-\xi_0}{\sqrt{(c^2-1)}}\right) \right\}_t + \left\{ \frac{c^2+1}{c^2-1} A - \frac{2Ac^2}{c^2-1} \cos^2\left(\frac{\xi-\xi_0}{\sqrt{(c^2-1)}}\right) \right\}_x = 0 \tag{5.22f}$$

which, on averaging over a wavelength, yield

$$\frac{\partial}{\partial t}\left(\frac{Ac^2}{\sqrt{(c^2-1)}}\right) + \frac{\partial}{\partial x}\left(\frac{Ac}{c^2-1}\right) = 0 \tag{5.23a}$$

and

$$\frac{\partial}{\partial t}\left(\frac{Ac}{c^2-1}\right) + \frac{\partial}{\partial x}\left(\frac{A}{c^2-1}\right) = 0. \tag{5.23b}$$

Since we are interested in the variation of the wave number k expressed exclusively in terms of c by (5.20) and the amplitude $a = 2A$, we further reduce the above equations to get

$$a^2\left\{ kk_t + \frac{1}{2}\frac{k_x(2k^2+1)}{\sqrt{(k^2+1)}} \right\} + a(k^2+1)\left\{ a_t + \frac{k}{\sqrt{(k^2+1)}}a_x \right\} = 0$$

and

$$a\left\{ a_t + \frac{k}{\sqrt{(k^2+1)}}a_x \right\} + a^2\left\{ \frac{2k^2+1}{2k(k^2+1)}k_t + \frac{k_x}{\sqrt{(k^2+1)}} \right\} = 0$$

which, on properly combining them, yield

$$k_t + \omega'(k)k_x = 0 \tag{5.24a}$$

and

$$a_t + \omega'(k)a_x + \frac{a}{2}\omega''(k)k_x = 0 \tag{5.24b}$$

in view of the dispersion relation (5.21).

These equations admit a double characteristic

$$\frac{\mathrm{d}x}{\mathrm{d}t} = \omega'(k) \qquad\qquad (5.25)$$

along which

$$k = \text{constant} \qquad\qquad (5.26a)$$

and

$$\frac{\mathrm{d}a}{\mathrm{d}t} = -\frac{1}{2}\omega''(k)k_x a. \qquad\qquad (5.26b)$$

Thus in case of the *linear* equation (5.14) the system determining the variations in k and a is *parabolic* and there is only one characteristic speed which is equal to the group velocity.

The general solution of (5.24a) has the form $x - \omega'(k)t = f(k)$ where f is an arbitrary function. Writing this in the form

$$\frac{x}{t} - \omega'(k) = \frac{1}{t}f(k)$$

we find that for large values of x and t

$$x = \omega'(k)t \qquad\qquad (5.27)$$

and then eqn (5.26b) shows that as we move with the group velocity, the amplitude varies as

$$a = \frac{a_0}{t^{1/2}}. \qquad\qquad (5.28)$$

Therefore the amplitude decays inversely as the square root of t, which conforms with the general result obtained in Chapter 1.

We shall now determine the velocity of propagation of energy. Let Δx be distance between two neighbouring characteristics corresponding to k and $k + \mathrm{d}k$. Since the energy is proportional to the square of the amplitude, the time-rate of change of energy between two characteristics is proportional to

$$\frac{\mathrm{d}}{\mathrm{d}t}(a^2\Delta x) = 2a\Delta x\frac{\mathrm{d}a}{\mathrm{d}t} + a^2\frac{\mathrm{d}}{\mathrm{d}t}\Delta x$$

$$= O(\Delta x)^2 \qquad\qquad (5.29)$$

in view of the equations (5.25) and (5.26b).

Therefore the energy is also propagated at the group velocity.

This is a linear problem, we have just one characteristic speed, namely the group velocity.

5.3.2. *Nonlinear wave equation*

We shall now study the nonlinear equation

$$\phi_{tt} - \phi_{xx} + V'(\phi) = 0, \tag{5.30}$$

where $V'(\phi)$ is not a linear function as was assumed in §5.3.1.

As in the last section, we can easily derive the following conservation equations

$$[\tfrac{1}{2}\phi_t^2 + \tfrac{1}{2}\phi_x^2 + V]_t + [-\phi_t\phi_x]_x = 0 \tag{5.31a}$$

and

$$[-\phi_x\phi_t]_t + [\tfrac{1}{2}\phi_t^2 + \tfrac{1}{2}\phi_x^2 - V]_x = 0. \tag{5.31b}$$

Here the steady solution is given by

$$\phi = \Phi(\xi; c, A), \; \xi = x - ct \tag{5.32a}$$

where

$$\Phi_\xi = \left[\frac{2}{c^2 - 1} \{ A - V(\Phi) \} \right]^{1/2}. \tag{5.32b}$$

Assuming $c^2 > 1$, $A > V(\phi)$ we get

$$\xi = \sqrt{\left(\frac{c^2 - 1}{2} \right)} \int \frac{d\Phi}{\{ A - V(\Phi) \}^{1/2}} \tag{5.33a}$$

and

$$\lambda = \lambda(c, A) = \sqrt{\{ 2(c^2 - 1) \}} \int_{\Phi_1}^{\Phi_2} \frac{d\Phi}{[A - V(\Phi)]^{1/2}} \tag{5.33b}$$

where Φ_1 and Φ_2 are the zeros of

$$A - V(\Phi) = 0 \tag{5.34}$$

between which the lhs of the last equation is positive.

Following the general theory discussed in §5.2, we introduce

$$W(c, A) = (c^2 - 1) \oint \Phi_\xi \, d\Phi \tag{5.35a}$$

$$= \sqrt{\{ 2(c^2 - 1) \}} \oint \{ A - V(\Phi) \}^{1/2} \, d\Phi \tag{5.35b}$$

$$= \sqrt{(c^2 - 1)} \, G(A) \tag{5.35c}$$

where

$$G(A) = \sqrt{2} \oint \{ A - V(\Phi) \}^{1/2} \, d\Phi \tag{5.36}$$

depends only on A and is independent of c.

On differentiating with respect to A, we get

$$G'(A) = \frac{1}{\sqrt{2}} \oint \frac{d\Phi}{\{A - V(\Phi)\}^{1/2}}$$

$$= \frac{1}{(c^2 - 1)^{1/2}} \oint d\xi = \frac{\lambda}{\sqrt{(c^2 - 1)}} \qquad (5.37)$$

and

$$G''(A) = -\frac{1}{2\sqrt{2}} \oint \frac{d\Phi}{\{A - V(\Phi)\}^{3/2}}$$

$$= -\frac{1}{(c^2 - 1)^{3/2}} \int_0^\lambda \frac{d\xi}{\Phi_\xi^2} < 0. \qquad (5.38)$$

We can also write $G(A)$ as

$$G(A) = \int_0^\lambda (c^2 - 1)\Phi_\xi^2 \, d\xi > 0. \qquad (5.39)$$

On partially differentiating (5.35c) with respect to A, we have, by (5.37),

$$W_A = \sqrt{(c^2 - 1)} G'(A) = \lambda(c, A). \qquad (5.40)$$

Then we define the wave number k by the relation

$$k = \frac{1}{\lambda} = \frac{1}{W_A} \quad \text{or} \quad kW_A = 1, \qquad (5.41)$$

where we have, for convenience, dropped a factor of 2π.

Once we have defined the wavelength λ we may directly evaluate the mean values $\overline{p(\Phi)}$ by integrating over the wavelength:

$$\overline{p(\Phi)} = \frac{1}{\lambda} \int_0^\lambda p(\Phi) \, d\xi = k \oint \frac{p(\Phi)}{\Phi_\xi} \, d\Phi. \qquad (5.42)$$

In view of the above definition of the mean value, we have

$$\overline{\tfrac{1}{2}\phi_t^2 + \tfrac{1}{2}\phi_x^2} = \overline{\tfrac{1}{2}(c^2 + 1)\Phi_\xi^2} = \frac{1}{2}k\frac{c^2 + 1}{c^2 - 1}W \qquad (5.43a)$$

$$\overline{-\phi_t\phi_x} = \overline{c\Phi_\xi^2} = \frac{kc}{c^2 - 1}W \qquad (5.43b)$$

and

$$\overline{V(\Phi)} = \overline{A - \tfrac{1}{2}(c^2 - 1)\Phi_\xi^2} = A - \tfrac{1}{2}kW. \qquad (5.43c)$$

On differentiating (5.35b) partially with respect to c, we have

$$W_c = \frac{c}{c^2 - 1}W. \qquad (5.44)$$

Substituting the above averaged expressions in the conservation equation, we have the following averaged conservation equation

$$\frac{\partial}{\partial t}\{k(cW_c + AW_A - W)\} + \frac{\partial}{\partial x}\{kc(cW_c + AW_A - W) - cA\} = 0 \qquad (5.45)$$

and

$$\frac{\partial}{\partial t}(kW_c) + \frac{\partial}{\partial x}(ckW_c - A) = 0, \qquad (5.46)$$

where, in writing the second term in eqn (5.45), we have added $kcAW_A$ and subtracted its equivalent expression cA.

On performing the differentiations and collecting the coefficients of c, A, and W, we have

$$c\left\{\frac{\partial}{\partial t}(kW_c) + \frac{\partial}{\partial x}(ckW_c - A)\right\} + A\left\{\frac{\partial}{\partial t}(kW_A) + \frac{\partial}{\partial x}(ckW_A - c)\right\}$$

$$- W\left\{\frac{\partial k}{\partial t} + \frac{\partial}{\partial x}(kc)\right\} = 0$$

or
$$k_t + (kc)_x = 0 \qquad (5.47)$$

since the coefficient of c is zero in view of (5.46) and the coefficient of A is zero, since $kW_A = 1$.

Defining the frequency in terms of the wave number k and wave velocity c, we have

$$\omega = kc$$

so that eqn (5.47) can be written as

$$k_t + \omega_x = 0 \quad \text{or} \quad k_t + \omega'(k)k_x = 0. \qquad (5.48)$$

Eqn (5.48) is extremely important because we have established the kinematic relation $k_t + \omega_x = 0$ entirely on the basis of the averaging procedure. Moreover, along the characteristic

$$\frac{\mathrm{d}x}{\mathrm{d}t} = \omega'(k)$$

k (and consequently ω) is conserved.

We can conveniently write the two independent averaged conservation equations as

$$\frac{DW_A}{Dt} - W_A\frac{\partial c}{\partial x} = 0 \qquad (5.49)$$

and

$$\frac{DW_c}{Dt} - W_A \frac{\partial A}{\partial x} = 0, \tag{5.50}$$

where

$$\frac{D}{Dt} \equiv \frac{\partial}{\partial t} + c \frac{\partial}{\partial x}. \tag{5.51}$$

On expressing the above equation in terms of A and c with the help of the relations (5.40) and (5.44) (after substituting for W in terms of G), we have

$$G''A_t + cG''A_x + \frac{cG'}{c^2 - 1}c_t + \frac{G'}{c^2 - 1}c_x = 0 \tag{5.52}$$

and

$$cG'A_t + G'A_x - \frac{G}{c^2 - 1}c_t - \frac{Gc}{c^2 - 1}c_x = 0 \tag{5.53}$$

which admit the following two characteristics

$$C_{\pm}: \quad \frac{dx}{dt} = \frac{1 \pm c\alpha}{c \pm \alpha}, \quad \alpha = \left(\frac{-GG''}{G'^2} \right)^{1/2} \tag{5.54}$$

with the compatibility relations

$$\left. \begin{array}{l} \dfrac{dc}{c^2 - 1} - \sqrt{\left(\dfrac{-G''}{G} \right)} dA = 0 \text{ along } C_+ \\[4mm] \dfrac{dc}{c^2 - 1} + \sqrt{\left(\dfrac{-G''}{G} \right)} dA = 0 \text{ along } C_-. \end{array} \right\} \tag{5.55}$$

and

These equations define two Riemann invariants

$$r = \int_{c_0}^{c} \frac{dc}{c^2 - 1} - \int_{A_0}^{A} (-G''/G)^{1/2} dA,$$

$$s = \int_{c_0}^{c} \frac{dc}{c^2 - 1} + \int_{A_0}^{A} (-G''/G)^{1/2} dA. \tag{5.56}$$

Eqn (5.54) defines two characteristic speeds $\dfrac{1 \pm c\alpha}{c \pm \alpha}$. This fact has an important bearing on the nature of the solution. Let us consider, for example, a wave-train which is initially uniform with $A = A_0$ and $c = c_0$ outside some finite region. After some interaction period, the disturbance separates into two simple waves separated by a domain of constant values of A and c. In one simple wave C_+ characteristics are straight lines which carry constant values of r and where the other Riemann invariant is everywhere constant. In the second simple wave r has same value everywhere and s is constant along C_-

characteristics which are straight lines. Between these two simple waves c and A take up constant values. Since the wave number and amplitude are expressible in terms of c and A, the same qualitative statements are also applicable to them.

We shall now study the propagation of energy. The total energy of the system per unit length is given by

$$\tfrac{1}{2}\phi_t^2 + \tfrac{1}{2}\phi_x^2 + V(\phi) \tag{5.57}$$

which has the average value

$$k(cW_c + AW_A - W). \tag{5.58}$$

From the averaged conservation equation (5.45), the average flux of energy is given by

$$kc(cW_c + AW_A - W) - cA. \tag{5.59}$$

Therefore, the speed, at which the energy is propagated, is given by

$$\frac{kc(cW_c + AW_A - W) - cA}{k(cW_c + AW_A - W)} = \frac{cG}{(c^2 - 1)AG' + G}. \tag{5.60}$$

This is another important speed. Thus in the present nonlinear case there exist two characteristic speeds and an energy propagation speed, and we have to designate one of them as the group velocity. Alternately, we may say that the concept of 'group' which we discussed in Chapter 1 has to be modified and all the three speeds are equally important. Probably, we may still call the speed with which the energy is propagated as the group velocity.

5.4. The Korteweg–de Vries equation

The aim of this section is to determine the characteristic velocities of the modulation equations obtained from the KdV equation taken in the following form:

$$u_t + 6uu_x + u_{xxx} = 0. \tag{5.61}$$

To obtain its steady solutions, we substitute

$$\xi = x - ct \tag{5.62}$$

so that it reduces to the form

$$u_{\xi\xi\xi} = cu_\xi - 6uu_\xi \tag{5.63}$$

which on integration yields

$$u_{\xi\xi} = B + cu - 3u^2. \tag{5.64}$$

Multiplying the above equation by u_ξ and integrating, we have

$$\tfrac{1}{2}u_\xi^2 = -A + Bu + \tfrac{1}{2}cu^2 - u^3 = f(u) \tag{5.65}$$

where A and B are constants of integration.

In Chapter 2, we have seen that the general case in which (5.65) admits a bounded solution is the one in which $f(u)$ has three real zeros α, β, γ $(\alpha > \beta > \gamma)$ and $\beta < u < \alpha$. We note that eqn (5.65) contains three parameters c, A, and B and therefore we shall need three independent averaged conservation equations to determine the slow variations in them with respect to x and t. We shall first establish the three independent conservation equations.

The first equation is obtained by simply rearranging (5.61)

$$u_t + (3u^2 + u_{xx})_x = 0. \tag{5.66}$$

The second equation is obtained by multiplying (5.61) by u and rearranging the terms

$$(\tfrac{1}{2}u^2)_t + (2u^3 + uu_{xx} - \tfrac{1}{2}u_x^2)_x = 0. \tag{5.67}$$

The third conservation equation is obtained by multiplying (5.61) by $3u^2$ and rearranging the terms:

$$(u^3 - \tfrac{1}{2}u_x^2)_t + (\tfrac{9}{2}u^4 + 3u^2 u_{xx} + \tfrac{1}{2}u_{xx}^2 + u_x u_t)_x = 0. \tag{5.68}$$

The averaging process is simplified if we introduce the function $W(c, A, B)$ defined by

$$W(c, A, B) = -\oint u_\xi \, du$$

$$= -\sqrt{2} \oint (-A + Bu + \tfrac{1}{2}cu^2 - u^3)^{1/2} \, du, \tag{5.69}$$

where the integration has to be taken over the complete cycle, namely from β to α and then from α to β. As in §5.3.2, here the wavelength and wave number are defined as follows:

$$\lambda = \lambda(c, A, B) = \int_0^\lambda d\xi = \oint \frac{du}{u_\xi} = \frac{1}{\sqrt{2}} \oint \frac{du}{f^{1/2}(u)}$$

$$= W_A \tag{5.70}$$

by actual differentiation of (5.69) with respect to A and

$$k = k(c, A, B) = \frac{1}{\lambda(c, A, B)} = \frac{1}{W_A} \tag{5.71}$$

or

$$kW_A = 1.$$

We shall now find the averages of the field quantities over a wavelength as discussed in the general theory in §5.2. We thus have

$$\bar{u} = k \int_0^\lambda u\,\mathrm{d}\xi = k \oint \frac{u\,\mathrm{d}u}{u_\xi} = \frac{k}{\sqrt{2}} \oint \frac{u\,\mathrm{d}u}{f^{1/2}(u)}$$

$$= -kW_B, \tag{5.72}$$

$$\overline{3u^2 + u_{xx}} = \overline{3u^2 + u_{\xi\xi}} = \overline{B + cu} = B - ckW_B. \tag{5.73}$$

$$\overline{\tfrac{1}{2}u^2} = k \oint \frac{\tfrac{1}{2}u^2\,\mathrm{d}u}{u_\xi} = -kW_c, \tag{5.74}$$

$$\overline{2u^3 + uu_{xx} - \tfrac{1}{2}u_x^2} = \overline{2u^3 + uu_{\xi\xi} - \tfrac{1}{2}u_\xi^2}$$

$$= \overline{A + \tfrac{1}{2}cu^2}, \qquad \text{substituting for } u_{\xi\xi} \text{ and } u_\xi^2$$
$$\text{from (5.64) and (5.65)}$$

$$= A - kcW_c, \tag{5.75}$$

$$\overline{u^3 - \tfrac{1}{2}u_x^2} = \overline{u^3 - \tfrac{1}{2}u_\xi^2} = \overline{-u_\xi^2 - A + Bu + \tfrac{1}{2}cu^2}, \text{ from (5.65)}$$

$$= kW - A + B(-kW_B) + c(kW_c)$$

$$= -k(AW_A + BW_B + cW_c - W), \tag{5.76}$$

since
$$\overline{u_\xi^2} = k \oint u_\xi^2 \frac{\mathrm{d}u}{u_\xi} = k \oint u_\xi\,\mathrm{d}u = -kW \tag{5.77}$$

and
$$kW_A = 1,$$

and finally

$$\overline{\tfrac{9}{2}u^4 + 3u^2 u_{xx} + \tfrac{1}{2}u_{xx}^2 + u_x u_t}$$

$$= \overline{\tfrac{9}{2}u^4 + 3u^2 u_{\xi\xi} + \tfrac{1}{2}u_{\xi\xi}^2 - cu_\xi^2}$$

$$= \overline{\tfrac{1}{2}B^2 + Bcu - cu_\xi^2 + \tfrac{1}{2}c^2u^2}, \qquad \text{on substituting for } u_{\xi\xi} \text{ from (5.64)}$$

$$= \tfrac{1}{2}B^2 + Bc(-kW_B) - c(-kW) + c^2(-kW_c)$$

$$= \tfrac{1}{2}B^2 + Ac - kc(AW_A + BW_B + cW_c - W). \tag{5.78}$$

Substituting these averages in the conservation equations, we have the following averaged conservation equations:

$$(kW_B)_t + (ckW_B - B)_x = 0 \tag{5.79}$$

$$(kW_c)_t + (ckW_c - A)_x = 0 \tag{5.80}$$

and

$$[k(AW_A + BW_B + cW_c - W)]_t$$
$$+ [ck(AW_A + BW_B + cW_c - W) - \tfrac{1}{2}B^2 - Ac]_x = 0. \qquad (5.81)$$

As in §5.3.2, we expand (5.81) keeping kW_A, kW_B and kW_c together and collect the terms in A, B, c, separately. The resultant equation reduces to

$$k_t + (ck)_x = 0 \qquad (5.82)$$

which is the conservation law for the wave number derived here entirely following the averaging procedure. Eqn (5.82) in terms of W_A takes the form:

$$\frac{DW_A}{dt} - W_A \frac{\partial c}{\partial x} = 0 \qquad (5.83)$$

and eqns (5.79) and (5.80) can be written as

$$\frac{DW_B}{Dt} - W_A \frac{\partial B}{\partial x} = 0 \qquad (5.84)$$

and

$$\frac{DW_c}{Dt} - W_A \frac{\partial A}{\partial x} = 0, \qquad (5.85)$$

where

$$\frac{D}{Dt} \equiv \frac{\partial}{\partial t} + c \frac{\partial}{\partial x}. \qquad (5.86)$$

Eqns (5.83)–(5.85) are our basic equations for studying the slow evolution of A, B, c. This information determines the ultimate behaviour of $u = u(\xi)$, where from (5.65), we have

$$\xi = \frac{1}{\sqrt{2}} \int \frac{du}{[-A + Bu + \tfrac{1}{2}cu^2 - u^3]^{1/2}}. \qquad (5.87)$$

Though the set of equations (5.83)–(5.85) appears simple, it is very tedious to determine the characteristic speeds from them. We find it easier to work through the zeros α, β, γ of $f(u)$ than to work with A, B, c.

From the relations between the roots and the coefficients of the cubic equations $f(u) = 0$, we have

$$\left. \begin{array}{l} \alpha + \beta + \gamma = \tfrac{1}{2}c \\[4pt] \alpha\beta + \beta\gamma + \gamma\alpha = -B \\[4pt] \alpha\beta\gamma \qquad = -A. \end{array} \right\} \qquad (5.88)$$

and

In terms of α, β, γ, we can write W as

$$W = -2\sqrt{2} \int_{\beta}^{\alpha} [(\alpha - u)(u - \beta)(u - \gamma)]^{1/2} \, du \qquad (5.89)$$

so that

$$W_{\alpha} = \frac{2\sqrt{\{2(\alpha - \gamma)\}}}{3} \{(\beta + \gamma - 2\alpha)E + (\beta - \gamma)K\} \qquad (5.90a)$$

$$W_{\beta} = \frac{2\sqrt{\{2(\alpha - \gamma)\}}}{3} \{(\alpha + \gamma - 2\beta)E + (\beta - \gamma)K\} \qquad (5.90b)$$

and

$$W_{\gamma} = \frac{2\sqrt{\{2(\alpha - \gamma)\}}}{3} \{(\alpha + \beta - 2\gamma)E - 2(\beta - \gamma)K\}, \qquad (5.90c)$$

where E and K are the complete elliptic integrals

$$E(s^2) = \int_0^{\pi/2} \Lambda \, d\theta, \qquad K(s^2) = \int_0^{\pi/2} \frac{d\theta}{\Lambda},$$

$$\Lambda = (1 - s^2 \sin^2 \theta)^{1/2} \qquad (5.91)$$

and the argument

$$s^2 = \frac{\alpha - \beta}{\alpha - \gamma} < 1 \qquad (5.92)$$

In establishing the above relations, we have used the following

$$H(s^2) = \int_0^{\pi/2} \Lambda^3 \, d\theta = \frac{2(\alpha + \beta - 2\gamma)}{3(\alpha - \gamma)} E(s^2) - \frac{\beta - \gamma}{3(\alpha - \gamma)} K(s^2). \qquad (5.93)$$

From (5.90), we have the important relation

$$W_{\alpha} + W_{\beta} + W_{\gamma} = 0. \qquad (5.94)$$

Now from (5.88), we have

$$d\alpha = \frac{1}{\Delta}(\beta - \gamma)(-dA + \alpha dB + \tfrac{1}{2}\alpha^2 dc) \left.\begin{array}{c} \\ \\ \\ \end{array}\right\}$$

$$d\beta = \frac{1}{\Delta}(\gamma - \alpha)(-dA + \beta dB + \tfrac{1}{2}\beta^2 dc) \qquad (5.95)$$

and

$$d\gamma = \frac{1}{\Delta}(\alpha - \beta)(-dA + \gamma dB + \tfrac{1}{2}\gamma^2 dc)$$

where

$$\Delta = (\alpha - \beta)(\beta - \gamma)(\alpha - \gamma). \tag{5.96}$$

The above relations determine partial derivatives of α, β, γ with respect to c, A, and B. Then we can express the partial derivatives of W with respect to c, A, B in terms of the partial derivatives of W with respect to α, β, γ

$$
\left.
\begin{aligned}
W_A &= 2\sqrt{\left(\frac{2}{\alpha - \gamma}\right)}K \\
W_B &= -2\sqrt{\left(\frac{2}{\alpha - \gamma}\right)}\{\gamma K + (\alpha - \gamma)E\} \\
W_c &= -\tfrac{1}{3}\sqrt{\left(\frac{2}{\alpha - \gamma}\right)}\big[\{\gamma(\alpha + \beta + 2\gamma) - \alpha\beta\}K \\
&\qquad + 2(\alpha - \gamma)(\alpha + \beta + \gamma)E\big]
\end{aligned}
\right\}
\tag{5.97}
$$

and

where we have suppressed the argument s^2 of E and K for sake of convenience of writing.

Now denoting by superior dot ($\cdot$) the derivative $\dfrac{D}{Dt}$, performing the operation on (5.97), and using

$$\dot{E} = \frac{1}{2}\frac{(\dot{s^2})}{s^2}(E - K) \tag{5.98a}$$

$$\dot{K} = \frac{1}{2}\frac{(\dot{s^2})}{s^2}\left(\frac{E}{1 - s^2} - K\right) \tag{5.98b}$$

and

$$\frac{(\dot{s^2})}{s^2} = \frac{(\beta - \gamma)\dot{\alpha} - (\alpha - \gamma)\dot{\beta} + (\alpha - \beta)\dot{\gamma}}{2(\alpha - \beta)(\alpha - \gamma)}, \tag{5.99}$$

we have

$$\dot{W}_A = \sqrt{\left(\frac{2}{\alpha - \gamma}\right)}\left[\frac{E - K}{\alpha - \beta}\dot{\alpha} + \frac{(\beta - \gamma)K - (\alpha - \gamma)E}{(\alpha - \beta)(\beta - \gamma)}\dot{\beta} + \frac{E}{\beta - \gamma}\dot{\gamma}\right] \tag{5.100a}$$

$$\dot{W}_B = -\sqrt{\left(\frac{2}{\alpha - \gamma}\right)}\left[\frac{\alpha E - \beta K}{\alpha - \beta}\dot{\alpha} + \frac{\alpha(\beta - \gamma)K - \beta(\alpha - \gamma)E}{(\alpha - \beta)(\beta - \gamma)}\dot{\beta}\right. \tag{5.100b}$$
$$\left. + \frac{(\beta - \gamma)K + \gamma E}{\beta - \gamma}\dot{\gamma}\right]$$

and

$$\dot{W}_c = -\sqrt{\left(\frac{2}{\alpha - \gamma}\right)}\left(\frac{\dot{\alpha}}{2(\alpha - \beta)}\left[-(\alpha\beta + \beta\gamma - \gamma\alpha)K \right.\right.$$

$$+ \{\beta\gamma - \alpha(\beta + \gamma - 2\alpha)\}E] + \frac{\dot{\beta}}{2(\alpha - \beta)}\left[(\alpha\beta + \alpha\gamma - \beta\gamma)K\right.$$

$$\left.- \frac{\alpha - \gamma}{\beta - \gamma}\{\alpha\gamma - \beta(\alpha + \gamma - 2\beta)\}E\right]$$

$$\left.+ \frac{\dot{\gamma}}{2(\beta - \gamma)}[2\gamma(\beta - \gamma)K + \{\alpha\beta - \gamma(\alpha + \beta - 2\gamma)\}E]\right). \qquad (5.100c)$$

We are now in a position to write the average conservation equations in terms of derivatives of α, β, and γ

$$(W_\gamma - W_\beta)\dot{\alpha} + (W_\alpha - W_\gamma)\dot{\beta} + (W_\beta - W_\alpha)\dot{\gamma}$$

$$= 4W_A \Delta(\alpha_x + \beta_x + \gamma_x) \qquad (5.101)$$

$$\{(\gamma - 2\beta)W_\gamma - (\beta - 2\gamma)W_\beta\}\dot{\alpha} + \{(\alpha - 2\gamma)W_\alpha$$

$$- (\gamma - 2\alpha)W_\gamma\}\dot{\beta} + \{(\beta - 2\alpha)W_\beta - (\alpha - 2\beta)W_\alpha\}\dot{\gamma}$$

$$= -2W_A \Delta\{(\beta + \gamma)\alpha_x + (\alpha + \gamma)\beta_x + (\alpha + \beta)\gamma_x\} \qquad (5.102)$$

and

$$\{(3\alpha\beta - 3\alpha\gamma + \beta\gamma)W_\gamma - (3\alpha\gamma - 3\alpha\beta + \beta\gamma)W_\beta\}\dot{\alpha}$$

$$+ \{(3\beta\gamma - 3\beta\alpha + \gamma\alpha)W_\alpha - (3\beta\alpha - 3\beta\gamma + \gamma\alpha)W_\gamma\}\dot{\beta}$$

$$+ \{(3\gamma\alpha - 3\gamma\beta + \alpha\beta)W_\beta - (3\gamma\beta - 3\gamma\alpha + \alpha\beta)W_\alpha\}\dot{\gamma}$$

$$= 4W_A \Delta\{\beta\gamma\alpha_x + \alpha\gamma\beta_x + \alpha\beta\gamma_x\}. \qquad (5.103)$$

We note that the above set of equations is invariant under cyclic permutation of α, β, γ.

To reduce these equations in the characteristic form, we multiply (5.101) by λ, (5.102) by μ and add them to (5.103). If we now choose λ and μ such that the coefficients of $\dot{\alpha}$ and α_x are zero, we get

$$\lambda = \alpha\beta + \gamma\alpha - 3\beta\gamma \qquad \text{and} \qquad \mu = 2\alpha. \qquad (5.104)$$

With these values of λ and μ, we find that the coefficients of β_x and γ_x are both equal to $-4W_A \Delta(\alpha^2 - \alpha\beta - \alpha\gamma + \beta\gamma)$, while the coefficients of $\dot{\beta}$ and $\dot{\gamma}$ are both equal to

$$2(\alpha^2 - \alpha\beta - \alpha\gamma + \beta\gamma)(W_\gamma - W_\beta),$$

where we have used the relation (5.104) in simplification. Thus the modified set reduces to

$$\dot{\beta} + \dot{\gamma} = \frac{2W_A \Delta}{W_\beta - W_\gamma}(\beta_x + \gamma_x). \qquad (5.105)$$

In view of the invariance under cyclic permutation of α, β, γ, we get other two equations by cyclically permuting α, β, γ.

From (5.105) and two similar expressions we get the following characteristic speeds:

$$c - \frac{2W_A \Delta}{W_\beta - W_\gamma} = c - \frac{4aK}{K - E} \qquad (5.106a)$$

$$c - \frac{2W_A \Delta}{W_\gamma - W_\alpha} = c - \frac{4aK(1 - s^2)}{E - (1 - s^2)K} \qquad (5.106b)$$

and

$$c - \frac{2W_A \Delta}{W_\alpha - W_\beta} = c + \frac{4a(1 - s^2)}{s^2}\frac{K}{E}, \qquad (5.106c)$$

where

$$a = \frac{\alpha - \beta}{2}.$$

The corresponding compatibility relations are $\beta + \gamma = $ constant, $\gamma + \alpha = $ constant and $\alpha + \beta = $ constant.

The wave number k is given by

$$k = \frac{1}{W_A} = \frac{a^{1/2}}{2sK}$$

so that we can write

$$s = s(a/k^2) \qquad (5.107)$$

and the three propagation speeds as

$$c + F(a/k^2) \qquad (5.108)$$

where F depends on a and k through the combination a/k^2.

Therefore the limit $a/k^2 \to 0$ corresponds to the linear case and should reproduce the results for the linearized form of the KdV equation

$$u_t + u_{xxx} = 0. \qquad (5.109)$$

The Fourier component corresponding to

$$u \propto \exp\{2\pi i(kx - \omega t)\} \qquad (4.110)$$

yields the following recurring relation for (5.109):

$$\omega = -4\pi^2 k^3.$$

(5.111)

In the limits $\dfrac{a}{k^2} \to 0$, the propagation speeds (5.106) reduce to

$$-3(2\pi k)^2, \qquad -3(2\pi k)^2, 0.$$

(5.112)

Clearly the first two speeds in (5.112) are equal to the group velocity given by eqn (5.111). The appearance of the third velocity, namely 0, in (5.112) can be explained as follows. The uniform solution of (5.109) is

$$u(x, t) = b + a \sin\{2\pi k x - (2\pi k)^3 t\}$$

(5.113)

and the average equations of the linear problem give $\dfrac{\partial b}{\partial t} = 0$ in accordance with the zero velocity in (5.112). However, we have to remember that the addition of a constant mean value to the solution is usually neglected as trivial in a linear theory.

5.5. Group velocity: dynamical treatment

Let us consider a general continuum system. Let $\eta_i (i = 1, 2, \ldots, n)$ be local variables or state variables and L be the Lagrangian density, i.e., Lagrangian per unit volume. We consider a system where L is a function of η_i and their first derivatives

$$\dot{\eta}_i \equiv \frac{\partial}{\partial t}\eta_i, \quad \eta_i^\alpha \equiv \frac{\partial}{\partial x_\alpha}\eta_i$$

(5.114)

which do not contain the independent variables i.e. space coordinates (x_1, x_2, x_3) and time coordinate t. Thus we have

$$L = L(\eta_i, \dot{\eta}_i, \eta_i^\alpha).$$

(5.115)

The Hamiltonian principle determines the time of evolution of the system, which states that 'the time-integral of the Lagrangian is stationary', which in the usual mathematical notation is written as

$$\delta \int_{t_1}^{t_2} dt \int_V L \, dV = 0$$

(5.116)

for any changes $\delta\eta_i$ in the functions $\eta_i(x_\alpha, t)$ which vanish at the beginning $t = t_1$ and at the end $t = t_2$ of an arbitrary time interval and also on the boundary of the arbitrary volume V of the three dimensional space (x_1, x_2, x_3) over which the integral in (5.116) is evaluated.

(5.116) yields the Lagrangian equations

$$\frac{\partial}{\partial t}\left(\frac{\partial L}{\partial \dot{\eta}_i}\right) + \sum_{\alpha=1}^{3} \frac{\partial}{\partial x_\alpha}\left(\frac{\partial L}{\partial \eta_i^\alpha}\right) - \frac{\partial L}{\partial \eta_i} = 0,$$

(5.117)

$$i = 1, 2, \ldots, n.$$

The total energy density of the system at (x_α, t) is given by (Goldstein 1950)

$$E = \sum_{i=1}^{n} \dot{\eta}_i \frac{\partial L}{\partial \dot{\eta}_i} - L \tag{5.118}$$

so that, in view of (5.117), we have

$$\frac{\partial E}{\partial t} = - \sum_{\alpha=1}^{3} \frac{\partial I_\alpha}{\partial x_\alpha}, \tag{5.119}$$

where

$$I_\alpha = \sum_{i=1}^{n} \dot{\eta}_i \frac{\partial L}{\partial \eta_i^\alpha}. \tag{5.120}$$

From (5.119), it is clear that the vector $\mathbf{I} = (I_1, I_2, I_3)$ represents the energy flux. It also tells us that the energy in a rectangular element with faces parallel to the coordinate planes changes at a rate equal to the difference between the energy flux across the opposite faces of the element.

For periodic plane waves, we may define the energy propagation velocity u_α as

$$u_\alpha = \frac{\langle I_\alpha \rangle}{\langle E \rangle} = \frac{\left\langle \sum\limits_{i=1}^{n} \dot{\eta}_i \dfrac{\partial L}{\partial \eta_i^\alpha} \right\rangle}{\left\langle \sum\limits_{i=1}^{n} \dot{\eta}_i \dfrac{\partial L}{\partial \dot{\eta}_i} - L \right\rangle}, \tag{5.121}$$

where $\langle E \rangle$ denotes average of the energy density obtained by averaging it over an integral number of wavelengths or periods. In terms of u_α, the averaged form of (5.119) takes the form

$$\frac{\partial}{\partial t}\langle E \rangle + \sum_{\alpha=1}^{3} \frac{\partial}{\partial x_\alpha}(\langle E \rangle u_\alpha) = 0. \tag{5.122}$$

If we are interested only in the plane periodic waves, we can use Hamilton's principle in the following slightly different form

A plane periodic wave satisfies

$$\delta \int_{t_1}^{t_2} \mathrm{d}t \int_V L \, \mathrm{d}V = 0 \tag{5.123}$$

for all changes $\delta\eta_i$ that are periodic with the same frequency and wave number as the η_i themselves provided that $t_2 - t_1$ is an integral multiple of the period and V is the rectangular volume with four of its faces perpendicular to the wavefront and integral number of wavelengths apart.

The proof of the above principle is more or less immediate as seen below, we have

$$\delta \int_{t_1}^{t_2} dt \int_V L \, dV = \int_{t_1}^{t_2} dt \int_V dV \sum_{i=1}^{n} \left(\frac{\partial L}{\partial \eta_i} \delta \eta_i + \frac{\partial L}{\partial \dot{\eta}_i} \delta \dot{\eta}_i + \sum_{\alpha=1}^{3} \frac{\partial L}{\partial \eta_i^\alpha} \delta \eta_i^\alpha \right)$$

$$= \int_{t_1}^{t_2} dt \int_V dV \sum_{i=1}^{n} \left\{ \frac{\partial L}{\partial \eta_i} - \frac{\partial}{\partial t}\left(\frac{\partial L}{\partial \dot{\eta}_i} \right) - \sum_{\alpha=1}^{3} \frac{\partial}{\partial x_\alpha}\left(\frac{\partial L}{\partial \eta_i^\alpha} \right) \right\} \delta \eta_i$$

$$+ \int_V dV \int_{t_1}^{t_2} dt \frac{\partial}{\partial t}\left(\sum_{i=1}^{n} \frac{\partial L}{\partial \dot{\eta}_i} \delta \eta_i \right)$$

$$+ \int_{t_1}^{t_2} dt \int_V \sum_{\alpha=1}^{3} \frac{\partial}{\partial x_\alpha}\left(\sum_{i=1}^{n} \frac{\partial L}{\partial \eta_i^\alpha} \delta \eta_i \right) dV = 0.$$

The first term is zero in view of the Lagrangian equations (5.117); the second term is zero on integrating with respect to t and remembering that $t_2 - t_1$ is an integral multiple of the period; and the third term is zero on evaluating the triple integral as a surface integral using the Gauss divergence theorem and remembering that the net contributions from the opposite faces of V vanish, being separated by integral multiples of the wavelength, and that the surface integrals are taken along the outward normals to these faces.

In the linear case, i.e. in the case of infinitesimal amplitudes, we can show that $\langle L \rangle = 0$. First, let us take the classical dynamical case.

Here $L =$ kinetic energy $-$ potential energy, so that we have

$\langle L \rangle =$ mean kinetic energy $-$ mean potential energy $= 0$.

In a general system subject to Hamilton's principle for waves of infinitesimal amplitude, L is a homogeneous function of degree two in all variables occurring as its arguments. Therefore, if we change each η_i to $(1 + \varepsilon)\eta_i$, and so on, then the Lagrangian L becomes $(1 + \varepsilon)^2 L$, so that the above variational principle yields

$$0 = \delta \int_{t_1}^{t_2} dt \int_V L \, dV = (1 + \varepsilon)^2 \int_{t_1}^{t_2} dt \int_V L \, dV$$

and therefore

$$\int_{t_1}^{t_2} dt \int_V L \, dV = 0,$$

which means that $\langle L \rangle = 0$.

In the case of a nonlinear wave, L is not necessarily a homogeneous function of its arguments and consequently

$$\langle L \rangle \neq 0$$

in general for such a wave.

We now enquire if we can express the group velocity (u_α) as

$$u_\alpha = \frac{\partial \omega}{\partial k_\alpha} \tag{5.124}$$

even in the case of plane waves of finite amplitude just as we have in case of waves of infinitesimal amplitudes. We have seen in Chapter 2 that in the case of nonlinear waves ω is not only function of the wave number **k** but also of the parameters representing the amplitude and other parameters which have been called *pseudo-frequencies* by Whitham (1965). However, we shall not consider the pseudo-frequencies here. Thus the derivative in (5.124) can have meaning when it is understood in the sense *keeping constant the two components of* **k** *other than* k_α *and also keeping constant some measure of the amplitude.*

Whitham (1965) has shown that (5.124) holds if the two components of **k** other than k_α and $\langle L \rangle / \omega$ are held fixed, i.e.

$$u_\alpha = \left(\frac{\partial \omega}{\partial k_\alpha} \right)_{\langle L \rangle / \omega} \tag{5.125}$$

We note that the classical result for linear waves is a particular case of (5.125) since for such waves $\langle L \rangle / \omega = 0$.

We shall now prove (5.125) for a general case.

Let us consider the plane periodic waves in the form

$$\eta_i = f_i \left(\omega t - \sum_{\alpha=1}^{3} k_\alpha x_\alpha \right) \tag{5.126}$$

where functions $f_i(\xi)$ are all periodic functions of $\xi = \omega t - \sum_{\alpha=1}^{3} k_\alpha x_\alpha$ with period P. We shall take $P = 1$ without loss of generality as it simply means scaling of the variable ξ.

From (5.126), we have

$$\left. \begin{array}{l} \eta_i = f_i(\xi) \\ \dot{\eta}_i = f_i'(\xi)\omega \\ \eta_i^\alpha = -f_i'(\xi)k_\alpha. \end{array} \right\} \tag{5.127}$$

and

We consider a general perturbation in which even the frequency and the wave number change:

$$\eta_i + \delta\eta_i = F_i \left(\Omega t - \sum_{\alpha=1}^{3} K_\alpha x_\alpha \right) \tag{5.128}$$

$$\left.\begin{array}{c} \Omega = \omega + \delta\omega, \qquad K_\alpha = k_\alpha + \delta k_\alpha \\[2mm] F_i(\xi) = f_i(\xi) + \delta f_i(\xi). \end{array}\right\} \qquad (5.129)$$

and

where $\delta f_i(\xi)$ have period 1 just like $f_i(\xi)$.

Now we have

$$\delta\langle L \rangle = \delta \int_0^1 L\{f_i(\xi), \omega f_i'(\xi), -k_\alpha f_i'(\xi)\} \, d\xi$$

$$= \int_0^1 \sum_{i=1}^n \left(\frac{\partial L}{\partial \eta_i} \delta f_i + \frac{\partial L}{\partial \dot{\eta}_i} \omega \delta f_i' - \sum_{\alpha=1}^3 \frac{\partial L}{\partial \eta_i^\alpha} k_\alpha \delta f_i' \right) d\xi$$

$$+ \left(\int_0^1 \sum_{i=1}^n \frac{\partial L}{\partial \dot{\eta}_i} f_i' \, d\xi \right) \delta\omega - \sum_{\alpha=1}^3 \left(\int_0^1 \sum_{i=1}^n \frac{\partial L}{\partial \eta_i^\alpha} f_i' \, d\xi \right) \delta k_\alpha. \qquad (5.130)$$

We note that the first term on the right-hand side of (5.130) represents the contribution of the changes resulting from the changes of f_i without change of ω and k_α and must vanish in view of Hamilton's principle for plane periodic waves for changes η_i which maintain frequency and wave number unchanged, while the second term and the third term on multiplying by ω can be expressed in terms of the mean values that appear in (5.121). Thus we have

$$\omega\delta\langle L \rangle = \left(\int_0^1 \sum_{i=1}^n \frac{\partial L}{\partial \dot{\eta}_i} \dot{\eta}_i \, d\xi \right) \delta\omega - \sum_{\alpha=1}^3 \left(\int_0^1 \sum_{i=1}^n \frac{\partial L}{\partial \eta_i^\alpha} \dot{\eta}_i \, d\xi \right) \delta k_\alpha$$

$$= \langle \sum_{i=1}^n \frac{\partial L}{\partial \dot{\eta}_i} \dot{\eta}_i \rangle \delta\omega - \sum_{\alpha=1}^3 \langle \sum_{i=1}^n \frac{\partial L}{\partial \eta_i^\alpha} \dot{\eta}_i \rangle \delta k_\alpha. \qquad (5.131)$$

When $\langle L \rangle / \omega = $ constant, we have

$$\frac{\delta\langle L \rangle}{\omega} - \frac{\langle L \rangle \delta\omega}{\omega^2} = 0$$

or

$$\omega\delta\langle L \rangle = \langle L \rangle \delta\omega. \qquad (5.132)$$

Substituting (5.132) in (5.131), we have

$$\langle L \rangle \delta\omega = \langle \sum_{i=1}^n \dot{\eta}_i \frac{\partial L}{\partial \dot{\eta}_i} \rangle \delta\omega - \sum_{\alpha=1}^3 \langle \sum_{i=1}^n \dot{\eta}_i \frac{\partial L}{\partial \eta_i^\alpha} \rangle \delta k_\alpha$$

or

$$\delta\omega = \frac{\sum_{\alpha=1}^3 \left\langle \sum_{i=1}^n \dot{\eta}_i \frac{\partial L}{\partial \eta_i^\alpha} \right\rangle \delta k_\alpha}{\left\langle \sum_{i=1}^n \dot{\eta}_i \frac{\partial L}{\partial \dot{\eta}_i} \right\rangle - \langle L \rangle}$$

which with the help of (5.121) gives (5.125)

The quantity $\langle L \rangle / \omega$ which is kept fixed in (5.125) is the integral of the Lagrangian density with respect to time over a single period. It is the quantity which remains stationary, when we go from a periodic solution η_i to the neighbouring value $\eta_i + \delta\eta_i$ which are periodic with the same frequency and wave number as η_i but are not, in general, the solutions. This explains what appears to be surprising in the above proof, namely that the functions (5.128) with their perturbed frequencies and wave numbers, which are solutions of the equations of motion, have not been used in the proof. It was not necessary because $\langle L \rangle / \omega$ would be the same as for the neighbouring functions which are not solutions, i.e. we could prove the result for more general perturbations: $\eta_i + \delta\eta_i, \omega + \delta\omega, k_\alpha + \delta k_\alpha$ which are not the solutions of the equations of motion.

The result (5.125) established by Whitham (1965) is very outstanding indeed as it extends the usual formula for the group velocity for linear waves to the case of nonlinear waves provided the pseudo-frequencies are kept fixed.

We note that the above Lagrangian formulation permits us to go over to the three dimensional case in a natural and elegant manner.

Bibliography

GOLDSTEIN H., *Classical mechanics*. Addison-Wesley, Reading, Mass. (1950).

HAYES, W. D. Group velocity and nonlinear dispersive wave propagation. *Proc. R. Soc. A* **332**, 199–221 (1973).

LIGHTHILL, M. J. 1965: Group velocity. *J. Inst. Math. and its Appl.* **1**, 1–28 (1965).

WHITHAM, G. B. Nonlinear dispersive waves. *Proc. R. Soc. A* **283**, 238–61 (1965).

——— A general approach to linear and nonlinear dispersive waves using Lagrangian, *J. Fluid Mech.* **22**, 273–83 (1965).

———*Linear and nonlinear waves*, Wiley, New York (1974).

AUTHOR INDEX

Ablowitz, M. J., 88

Ames, W. F., 41

Asano, N., 39, 44, 49, 50, 54, 56

Bateman, H., 41

Bean, C. P., 39, 43

Benjamin, T. B., 38, 42

Benton, E. R., 41

Berezin, Y. A., 42

Bespalov, V. I., 39, 44

Bhatnagar, P. L., 59

de Blois, R. W., 39, 43

Bona, J. L., 38, 42

Born, M., 40, 44

Brown, G. H., 39, 43

Burgers, J. M., 42

Chu, C. W., 42

Chu, F. Y. E., 42

Cole, J. D., 42

Debye, 15

Dennery, P., 9, 14

Donth, H., 39, 43

Döring, W., 39, 43

Eddington, A. S., 6, 14

Enz, U., 39, 43

Feenberg, F., 40, 44

Fergason, J. L., 39, 43

Frenkel, J., 39, 43

Fulton, T. A., 39, 44

Gardner, C. S., 42, 66, 70, 88, 89, 92, 111

Gel'fand, I. M., 70, 88

de Gennes, P. G., 40, 44

Goldstein, H., 134, 138

Greene, J. M., 70, 88

Hahn, E. L., 40, 45

Hasegawa, A., 39, 44

Hayes, W. D., 138

Hirota, R., 39, 40, 43, 44

Hopf, E., 42

Ichikawa, V. H., 40, 44

Imamura, T., 39, 44

Infeld, L., 40, 44

Jeffrey, A., 42, 46, 50

Jeffreys, B., 9, 14

Jeffreys, H., 9, 14

Johnson, W. J., 39, 43

Kadomtsev, B. B., 42

Kakutani, T., 38, 42

Karpman, V. I., 39, 42, 44

Kay, I., 70, 72, 88

Kelley, P. L., 39, 44

Kochendörfer, A., 39, 43

Kontorova, T., 39, 43

Kruskal, E. M., 39, 44

Kruskal, M. D., 42, 66, 70, 88, 89, 92, 111

Krzywicki, A., 9, 14

Kulik, I. O., 39, 43

Kulikovskii, A. G., 59, 60

Lamb, G. L., 41, 45

Landau, L., 62, 72, 88

Lax, P. D., 91, 102, 111

Lebwohl, P., 39, 43

Leibovich, S., 42, 49, 50

Levitan, P. M., 70, 88

Lifschitz, E., 62, 72, 88

Lighthill, M. J., 11, 14, 31, 42, 138

Lusternik, L. A., 111

Mahony, J. J., 38, 42

McCall, S. L., 40, 45

Mc Laughlin, D. W., 42

Meleshko, V. P., 42

Miura, R. M., 42, 66, 70, 88, 89, 92, 111

Moses, H. E., 70, 72, 88

Newell, A. C., 88

Olsen, S. L., 40, 44

Ono, H., 38, 49, 50, 54, 56

Peregrine, D. H., 48, 50

Platzman, G. W., 41

Porter, J. R., 40, 45
Prasad, P., 39, 40, 43, 44, 49, 50, 59, 60
Price, J. D., 111

Ravindran, R., 39, 40, 43, 44, 49, 50
Rosen, N., 39, 43
Rosenstock, H. B., 39, 43

Scott, A. C., 39, 42, 43
Scott Russell, J., 38, 42
Seebass, R., 42, 49, 50
Seeger, A., 39, 43
Shimizu, K., 40, 44
Shvets, M. E., 42
Sjoberg, A., 42, 111
Skyrme, T. H. R., 39, 43
Slobodkina, F. A., 59, 60
Sobolev, V. I., 111

Stephen, M. J., 39, 43

Talanov, V. I., 39, 44
Tanaka, S., 88, 89
Taniuti, T., 39, 42, 44, 46, 49, 50, 54, 56
Tappert, F., 39, 44
Toda, M., 39, 43, 44, 88, 89

Varma, C. M., 39, 44

Wadati, M., 39, 43, 44, 88, 89
Washimi, H., 39, 44
Wei, C. C., 42, 49, 50, 54, 56
Whitham, G. B., 66, 115, 116, 117, 136, 138

Yajima, N., 39, 44

Zabusky, N. J., 38, 39, 42, 43
Zakharov, V. E., 88, 89

SUBJECT INDEX

Adjoint operator, 94
amplitude, 2, 10
antinodes, 4
antisymmetric operator, 95

Beats, 7
Bloch equation, 41
Born–Infeld equation, 40, 44
Boussinesq equation, 38, 43
Burgers equation, 21, 29, 41, 57, 58
  steady solution of, 29

Characteristics, 2
cnoidal wave, 35
commutator, 98
conservation law, 65, 115
crest, 2

Diffusive wave, 6, 22, 28, 31
dispersion, 8
dispersion relation, 4
dispersive wave, 6, 8, 22, 31
duct flow, 46, 57

Eigenvalues (*see* Schrödinger equation)
  eigenvalues of unitarily
  equivalent operators, 99
elliptic functions-Jacobian, 35
elliptic integrals, 35
energy, 11, 12, 134
  velocity of propagation of, 11, 120, 134
energy density, 11, 134
equations of evolution, 91
  integrals of, 100

Frechet derivative, 101
frequency, 3, 114, 136

Gel'fand–Levitan integral equation, 70
gradient of integral, 101
group velocity, 6, 113, 117, 133, 136

Hirota equation, 40, 44

Integral of equation, 65
inverse scattering problem, 70

Korteweg de Vries equation, 21, 32, 40, 42, 57, 61, 92, 104, 125
  associated, 66
  conservation form of, 65
  eigenspeeds of, 108, 109
  generalized form of, 38, 42
  integral of, 65
  one soliton solution of, 71
  regularized, 38
  two soliton solution of, 75
  N-soliton solution of, 80
  steady solution of, 32, 114

Lagrangian, 133
Lagrangian equations, 133
linear waves, 1
lattice equation, 39, 43

Mode, 5

Nodes, 4

Overstability, 6

Period, 3
phase, 1
phase function, 9
phase velocity, 2

Rankine–Hugoniot relation, 30
reduction theory, 51
reflection coefficient, 64, 69
Riemann invariants, 124

Saddle point method, 9, 15, 18
scattering parameters 64
Schrödinger equation, 61, 67
  continuous eigenvalues of, 62, 63, 88
  discrete eigenvalues of, 62, 67
  properties of, 62
  unsteady nonlinear, 39, 44
Schrödinger operator, 105
self-adjoint operator, 95
shallow water waves, 46, 48, 59
shocks, 25
  structure of, 30

thickness of, 31
sine Gordon equation, 39, 43
solitary wave, 37, 61, 100
soliton, 37, 38
soliton interaction, 61, 83
steepest descent, method of, 9, 15
symmetric operator, 95

Transmission coefficient, 64, 69
transparency equation, self-induced, 40, 45
trough, 2

Unitary equivalence, 94, 95
unitary operator, 95, 97

Variational equation, 101

Wave, 1
  Alfven, 38
  Cnoidal, 35
  dispersive, 6, 8
  diffusive, 6
  progressive, 1, 4
  standing, 4
  solitary (*see* solitary wave)
Wavelength, 3, 114
wave number, 3, 14, 114, 136
weak solution, 25, 28